Mathematics for Physical Geographers

Mathematics for Physical Geographers

Graham N. Sumner

A HALSTEAD PRESS BOOK

John Wiley & Sons
New York

First published in 1978 by Edward Arnold (Publishers) Limited
London

Published in the U.S.A
by Halsted Press, a Division of
John Wiley & Sons, Inc., New York.

Sumner, Graham N
Mathematics for physical geographers.
"A Halstead Press book."
Bibliography: p.
Includes index.
1. Physical geography—Mathematics. I. Title.
GB21.5.M33S95 1978 510 78–12156

ISBN 0–470–26557–4

Printed in Great Britain

Acknowledgements

The author and publishers wish to thank the following for permission to reproduce or modify copyright material:
W. H. Freeman and Co. for figure 5.12;
The Geological Society of London for figure 5.16;
Cambridge University Press for the tables in Appendix 3 and 4.

To Seirian Rose

Contents

Appendices

Preface

The author's thanks are due to his long-suffering research supervisor, Dr Maureen Jones, for critically reading the early versions of the chapters in this book, to Miss Angela Bannell and Mr Phil Bennie BA, who as students at Lampeter ensured that the message of the book reached those for whom it is intended, and to Dr D. A. Davidson who, as author of the companion volume to this book, provided many ideas and inspiration in the book's formative stages.

Introduction

Mathematics is of great importance to, and is frequently implicit within, the entire scientific approach. Physical geography must be regarded as a science which seeks to analyse relationships and establish statistical laws or scientific laws, which may be further used to predict future similar events or outcomes. Thus, physical geography is a branch of science which deals with change, the processes which induce change, and the speed with which these changes take place and processes operate. The use of field or laboratory experiments to establish relationships and the possible development of these relationships into scientific laws permits the outcome of similar events to be predicted to a known degree of accuracy. Mathematics provides the corner-stone by which observed relationships may be unambiguously portrayed and predictions made. Where we cannot be sure of the accuracy of our forecast as is often the case in physical geography, this prediction is itself dependent upon the operation of statistical laws, which in turn use the language of mathematics in their expression. The use and understanding of mathematics is thus fundamental to physical geography if we are to consider physical geography a 'true' science.

The statistical element to research in physical geography is an important one. Rarely are we able to establish a unique relationship between two parameters in physical geography: very often too much 'noise' creeps in to make the relationship less than certain for predictive purposes. This 'noise' emanates both from the influence of other parameters external to those under consideration and from an inadequacy and inaccuracy of instrumentation and measurement in the laboratory most familiar to physical geographers: the environment. Thus, when we are looking at the mathematics of *functions* (chapters 3 and 5) which are the mathematical expressions of relationships, we must inevitably draw upon an assumed awareness of certain statistical techniques and concepts on the part of the reader: this is not a statistical text. This is regretted but unavoidable. To the reader totally untutored in basic statistics a little help is available in the statistical appendix (Appendix 5), and in the references to certain, basic, statistical texts where appropriate.

Chapters 3 and 5 are probably the most important in the entire book. The establishment of mathematical functions to describe observed relationships is more than half the battle in constructing a mathematical model of the phenomena being studied. Once a function has been established we are able to investigate rates of change of one parameter with another by means of the *derivative* of the function (chapter 5), and we may further obtain an impression of the *cumulative* effect of one parameter upon another by taking the *integral* of the function (chapter 8). We have already observed though that our established relationships are frequently uncertain, and thus a knowledge of the mathematics of probability is an important piece of mathematical expertise which the physical geographer can scarce be without. This aspect is dealt with in chapter 7.

Very often however, we are as physical geographers, concerned with the modelling of entire *systems* in the environment. The most familiar of these systems in the physical geography texts of recent years have been the ecosystem and the drainage basin hydrological cycle. We may, for example, examine the progress of water through the latter system. Resultant upon precipitation input a number of different fates await water input to the system. The water will pass through the basin in a number of different ways, as water in rivers and streams, as subsurface flow or as overland flow. Each of these takes place at different rates, from the very fast (river discharge) to the very slow (molecular transfer of moisture within the soil) and represents the end result of a different process. All water ultimately forms an output from the basin, by evaporation, transpiration, loss to groundwater or simple discharge at the mouth of the basin. We can represent the behaviour of water within such a system mathematically. Over finite time periods we know that input must balance output or there will be progressive flooding or drying out of the basin. We have here a function whose individual elements describe the results of a different process; infiltration, open channel flow and so on. All are expressed as rates of water flow and represent a very simple type of *differential equation*. The solution of differential equations is dealt with in the last chapter of the book. The concept of a drainage basin also of course, entails the treatment of spatial variation. The depiction of variation in space through graphs is a simple mathematical matter. An extension of graph coordinates (chapter 1) to matrices (chapter 4) and through dynamic space (movement) makes an understanding of *vectors* (chapter 2) a further important part of the physical geographer's mathematical diet.

Of course mathematics is not the only support essential to the operation of physical geography as a science. The scientific approach and the use of the laws of physics and chemistry are equally important. Mathematics is also an important part of physics and chemistry, and all other related disciplines. Throughout this book the reader will find reference to the companion volume to this:

Science for Physical Geographers by D. A. Davidson.

Both texts have drawn upon the experience of attempting to teach physical geography at all levels to geography undergraduates drawn from dominantly Arts backgrounds, many of whom so often seem to have no more than a mental 'block' against science and mathematics, rather than any innate inability to comprehend the subject matter. Mathematics is no more difficult than the understanding of a language. It is just as possible to become fluent in mathematics as it is to become fluent in German or even Welsh! As with the learning of a language though, the various stages of comprehension must be acquired gradually from the most basic, through the elementary, to the advanced.

This text does not pretend to be advanced. It commences with the establishment of the most basic tenets of mathematics and culminates in an introduction to some of the more complicated aspects of calculus: integration and the solution of differential equations. These last two chapters might be regarded as the level to which physical geographers *ought* to aspire, but the previous seven represent the mathematical knowledge that the reader should regard as a *working minimum* for most present-day undergraduate geography courses. The first chapter lays down the foundations to arithmetic which are commonly taught in Primary schools! No apology is made for the inclusion of such a basic chapter since it is the author's experience that very many students have little grasp of this sort of background! Each subsequent chapter builds on the previous, and examples from physical geography are given to illustrate the use to which the various techniques may be put, although the major concern is with the acquisition of mathematical knowhow throughout the book. The book is aimed at first and second year honours geographers who may wish to specialize in physical geography. This is not to say that aspiring human geographers should ignore the book: in most cases the mathematical demands on human geographers are as great, and the mathematical knowledge required of them duplicates that required of physical geographers. The title of this book is conditioned by the interests of the author and by the need for such a text devoted to physical geography. It is the geographical content which determines the title, far less so the mathematical content.

Each chapter is accompanied by reference lists for further reading and by problems, the answers to most of which appear at the end of the book. Care has been taken in the selection of texts for further reading, so that the books chosen are particularly clearly written, often for non-mathematicians, or contain large numbers of further problems which the reader might wish to attempt in order to improve his expertise. Experience is the secret of ability in mathematics. Thus, great stress must be laid on the reader attempting *all* problems, and to work through them particularly carefully. If a wrong answer is obtained it is particularly important that the reason for error is found: the maxim of 'learning by one's mistakes'.

It is hoped that the book will fulfil a need in providing a *basic* mathematical text. No assumption is made of prior mathematical knowledge, not even a GCE Ordinary Level pass in the subject, even though the majority of undergraduate geographers should possess this! To those who need to learn (or relearn) basic mathematics as an aid to their geographical studies the book should be used as a teaching text and be worked through systematically. To others, who have perhaps recently passed GCE Advanced Level pure mathematics, an apology is made for the very basic approach, but it is hoped that even they will benefit from the book's use as a reference text. To both groups, a complete understanding of the mathematics contained in this book will permit the reading of the vast majority of research publications in physical geography, perhaps with the realization for the first time that the mathematical formulae and functions contained within them express far more concisely than the text itself, the points being made in the paper.

1

Mathematical conventions

The traditional threesome within elementary mathematics between arithmetic, algebra and geometry is still by and large the basis on which mathematical practice is built. Increasingly though the wider realm of subjects within mathematics, including calculus (chapters 5 and 8 in this book) and the use of vectors (chapter 2) for the solution of functions (chapters 3 and 5), has penetrated down from the exclusive domain of the pure mathematician through the pure sciences, and now forms an important part of the structure of research and teaching in physical geography. Thus there is sometimes a number of ways in which a problem may be solved mathematically. Ultimately though all mathematics stems from the very simple rules which govern the basic arithmetic, algebraic or geometric operations. The numbers used in arithmetic, or the general symbolization of algebra using letters from the English and Greek alphabets (see Appendix 1), and the use of spatial measures in geometry (section 2.8), are the basic building blocks of mathematics. Thus as well as rigorously defining what we mean by number and quantity, we also require a means by which these basic building blocks may be manipulated. The way in which we use them is indicated by symbols, and the mode of operation indicated by the symbol must be carefully defined and understood if we are to attain a rigorous and unambiguous framework within which to work.

At a general level mathematics shares much in common with music and literature in that within both these fields operations are indicated by the use of symbols whose meanings are strictly defined. The basic melody of a symphony is portrayed simply on paper by means of symbols indicating the time for which given notes sound, and the pitch of these notes, by their position relative to staves—akin to the use of the axes of a graph (section 1.8). The words in a novel and the letters from which the words are composed fulfil the same purpose, whilst grammar has its parallel in the harmony of the musical scales. The conventions laid down within mathematics are applied equally rigorously such that the framework provided is a rigid and workable one. The basic concepts of number (section 1.3) are similar to the musical notes or the letters of the alphabet and the written word. What is to be done with the numbers—how they are to be used—is indicated by the use of abbreviations common to all aspects of mathematics. Although the conventions governing their use are simple we are nevertheless able to express highly complicated and often abstract, concepts. Thus for example the relatively difficult solution of differential equations in chapter nine bears as much resemblance to the conventions laid down in this chapter as does Shostakovitch's tenth symphony to the scale of the key in which it is written, E minor.

1.1 Operations

Symbols used to indicate mode of operation in mathematics generally fall within two broad categories: those which *inform*, such as '=' indicating an equality of magnitude between two items, and those which *instruct*, such as '+' and '−' which are *operators*. Within the first category a number of different symbols are in common use:

$=$	equal to
$\neq$	not equal to
$\equiv$	equivalent to, representing
$\propto$	proportional to; varies as
$<$	less than ($\leq$ less than or equal to)
$>$	greater than ($\geq$ greater than or equal to)
$\nless$	not less than
$\ngtr$	not greater than
$\risingdotseq$ or $\simeq$ or $\doteq$	approximately equal to

Most symbols in the second group should be very familiar to all readers: $+$, $-$, $\times$, $\div$; add, subtract, multiply and divide, although there are variations which could confuse. The alternative representation of operations for multiplying or dividing two values a and b are given here:

a multiplied by b $a \times b$ or $a.b$ or ab

a divided by b $a \div b$ or $\dfrac{a}{b}$ or a/b or $a.b^{-1}$

Under SI (Système International) convention (National Physical Laboratory 1970) ab and a/b or ab^{-1} are preferred. As we shall see in chapter 2 the use of the point to indicate multiplication can usefully be confined to vector multiplication. Further symbols will be introduced when relevant throughout the ensuing chapters. There are however two very useful abbreviations which fall into neither of the above categories and which are often used in mathematics; the symbol $\therefore$ meaning 'therefore', and more rarely, the symbol $\because$ meaning 'because'.

1.2 Order of computation

Where several operations are to be carried out in a calculation the order in which successive computations are made can influence the result. In all composite computation such as $a+bc/d$, multiplication and division take precedent over addition and subtraction. Thus in the above expression bc/d would be computed first and the solution finally added to a, rather than a being added to bc and then the product divided by d. Where exceptions to this rule are required brackets are placed around the operation to be carried out first of all. Thus if we wrote $(a+b)c/d$, a must be added to b and then the solution multiplied by c/d. The result in each case is different. For example, if $a = 2$, $b = 6$, $c = 1$ and $d = 4$:

$$a+bc/d = 2+6/4 = 3{\cdot}5$$
$$(a+b)c/d = (2+6)1/4 = 2{\cdot}0$$

It is important to remember these rules when such operations are carried out using computers.

A further use of brackets is in the *distributive law of multiplication.* For example, the expression:

$$ac + ad + bc + bd$$

can be more conveniently represented as:

$$(a + b)(c + d)$$

by regrouping values within brackets. Here we have factorized the expression having noted that all four terms in the original expression have either a or b and either c or d in them. By grouping pairs of alternatives within two sets of brackets we arrive at the solution $(a + b)(c + d)$. Multiplying out, first $a(c + d)$ and then $b(c + d)$ we arrive at $ac + ad + bc + bd$, the original expression.

Very often in physical geography computation involves the use of both *positive* and *negative* values, often to indicate opposite senses, or to represent a range of values through zero from −n to +n. Thus, if say, f represents acceleration in an expression, $+f$ will indicate acceleration in one direction and $-f$, acceleration in the opposite direction. Both $-f$ and $+f$ have an absolute magnitude f, which may be written $|f|$: 'modulus f'. An important application of sign to indicate direction is found in the use of vectors (chapter 2), and Sumner (1977a) has used these in a study of sea breeze development. By defining all onshore gradient winds as *positive* and all offshore gradient winds as *negative* we can begin to assess the importance of gradient wind direction upon sea breeze development, given also the required clear skies and resultant shallow convection (see Davidson 1978, for a treatment of the processes involved). Clearly winds are only rarely directly onshore or directly offshore: that is blowing exactly at right-angles across the coastline in either direction. The impact of an onshore wind of v metres per second will clearly be greatest when the wind is directly onshore, and least when it is blowing more or less parallel to the coastline. The contribution such a wind makes as an onshore wind is reduced as its angle to the coastline is reduced (see figure 1.1). In fact, using one of the trigonometric ratios we shall establish in chapter 2, the onshore *component* of gradient wind is equal to $v \sin \alpha$ or $v \sin \beta$ in figure 1.1,

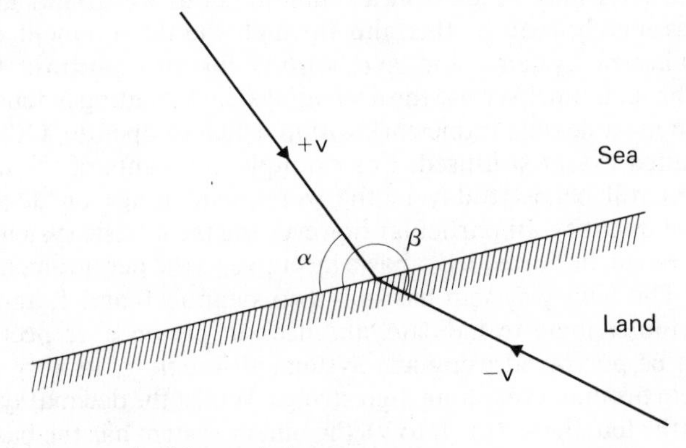

Figure 1.1 Onshore and offshore wind components.

and sin α is positive whilst sin β is negative (see table 2.1). The technique is further amplified later in chapter 2.

Computation with expressions including both positive and negative values involves additional rules. Except when + is used to indicate addition it is generally omitted unless it serves to clarify a contrast as in the sea breeze example above. In addition or subtraction operations a negative always 'cancels out' a positive, so that by subtracting $+b$ from $+a$ we obtain $a-b$,

$$(+a)-(+b) = a-b$$

and similarly,

$$(+a)+(-b) = a-b$$

With multiplication and division which are mathematically both the same process since $a/b = a(1/b)$, the rule is a little more complex. If an *even* number of negative values occurs then the result is always *positive*, thus:

$$(-a)(-b) = +ab$$

or
$$(-a)(-b)(-b)/(-c) = (-ab^2)/(-c) = +(ab^2/c)$$

For an *odd* number of negative values however the result is always *negative*, thus:

$$(-a)(+b) = -ab$$
$$(-a)(-b)/(-c) = +ab/-c = -(ab/c)$$
$$(-a)(-a)(-a) = -a^3$$

1.3 Quantity and number

So far we have restricted our attention to basic arithmetic operations at their most general. Algebraic representation has been used simply to indicate where numbers may be substituted, a or b in the preceding example could equally represent any feature or process to which magnitude may be ascribed, be it the amount of sediment in a river or the rate of evaporation. Quantity and measurement of it is an essential part of studies in physical geography. A general need to quantify has led to the development of certain conventions and concepts in order that we may formulate a rigid 'grammatical' framework within which data may be used in an unambiguous way. The 'language' of mathematics merely realizes this aim through the development of number systems. Different systems and symbolisms for this purpose have been developed through time. We use the *decimal* system (counting in tens), since for us this is the most flexible framework within which to operate. Other systems are to a limited extent still used. For example, remnants of the *duodecimal* system (12's) still persist today in the widespread usage of 'dozens' as an expression of quantity. In particular however the most basic system invented, the *binary* system, has provided a basic incentive to the development of digital computers. The binary system has only two symbols 0 and 1, and these are used in computer logic to indicate 'absence' or 'presence' respectively. Any number can be portrayed using any system, although by its very nature the binary system produces very long digit strings. Whilst the decimal system has a *base* of ten (the ten digits from 0 to 9), the binary system has the base two (the two digits 0 and 1). Within either of these numbering systems each digit has a prescribed value which permits the ordering of magnitude in the most elementary way. In addition larger (or smaller) magnitude numbers are defined by the *position* of further digits working in ascending order of magnitude from right to left. In the decimal system the first magnitude digit position (from right to left) is termed 'unit', the second 'ten', the third 'hundred' and so on. These are convenient names for *powers* of the base number ten—that is, the third order of magnitude (thousand) is 10^3, the second, 10^2 and the first 10^1. Thus 200 is 2×10^2, 500,000 is 5×10^5 and 40 is 4×10^1. Logically the progression of this sequence down from 10^1 indicates that 10^0 is one. Indeed the raising of *any* number to the power zero is always unity, or:

$$1^0 = 6^0 = 500{,}000^0 = 1$$

Thus the number 3,551 is a shorthand portrayal of:

$$(3 \times 10^3) + (5 \times 10^2) + (5 \times 10^1) + (1 \times 10^0)$$

A similar sequence can be shown for all number systems, so that under one using the base five (digits 0 to 4) the number 2,441 would be:

$$(2 \times 5^3) + (4 \times 5^2) + (4 \times 5^1) + (1 \times 5^0)$$

It is thus a relatively simple matter to convert numbers from one system to another. A simple worked example will assist in a greater understanding as well. The decimal number 53 can be converted into its binary equivalent by finding the highest power of two which will 'go' at least once into 53 (i.e. which is a denominator of 53). 2^5 has the value 32 whilst 2^6 has the value 64, so in table 1.1 by dividing 53 by 2^5 we obtain the remainder (column (d)) of 21. The highest power of two which is a denominator of 21 is the next part of the binary equivalent and so on. Ultimately a zero occurs in the remainder column and addition of powers of two in column (a) represents the binary equivalent. Thus the decimal 53 is expressed in binary form as:

$$(1 \times 2^5) + (1 \times 2^4) + (0 \times 2^3) + (1 \times 2^2) + (0 \times 2^1) + (1 \times 2^0)$$

or 110,101.

The systems as defined up to this point are sufficient merely to define whole or *integer* values (the *natural* numbers, which are any one of the counting integers 1, 2, 3, … ,*N*). *Real* numbers, for example the mathematical constant 3·141 2 (π): integer 3+0·141 2 (real), can be similarly expressed in terms of powers of the base, using decimal parts or fractions. To extend the notation to values of less than one is a simple matter of a continued reduction into negative powers. The traditional means of representing the beginning of negative

Table 1.1 Determination of the binary of decimal 53.

(a)	(b) value accounted	(c) accumulated total	(d) remainder
2^5	32	32	21
2^4	16	48	5
2^2	4	52	1
2^0	1	53	0

powers is the decimal point, thus 23·605 7 is merely a shorthand notation of:

$$(2\times 10^{1})+(3\times 10^{0})+(6\times 10^{-1})+(0\times 10^{-2})+(5\times 10^{-3})+(7\times 10^{-4})$$

Note that as with magnitude to the left of the decimal point, the individual numbers are grouped for convenience into blocks of three, but separated this time by spaces rather than commas. Magnitude is again reflected in the position of the digit in the string, so that just as 10^2 indicates 'one hundred times', 10^{-2} indicates 'one hundredth part', or as a fraction, 1/100; that is 10^{-2} is the *reciprocal* of 10^2 ($1/10^2$). A very common extension of this facility is to express numbers of very large or very small magnitude as ordinary decimals multiplied by a power of ten. For example, 3,456,000 may be rewritten '3·456 million' or as $3{\cdot}456\times 10^6$; similarly 0·000 034 can be written $3{\cdot}4\times 10^{-5}$.

1.4 Rational and irrational numbers

Many real numbers may be written in an even shorter form when expressed as fractions. The number 2·64 is thus $2\frac{64}{100}$ or $2\frac{16}{25}$. The number is termed *rational* since it is capable of exact representation as a fraction using two integer values. Many real numbers cannot be represented exactly using two integers however, and these are termed *irrational* numbers. Examples of such irrational numbers are π, 3·141 592 65 to eight decimal places, or $\sqrt{2}, \sqrt{3}$ and so on. However the fraction 2/9 is rational even though its decimal equivalent is 0·222 222 222 ... repeated over and over again. By representing the repeated or recurring part of the decimal with a dot placed over the repeated digits this can be rewritten $0{\cdot}\dot{2}$. It should be realized that although this number repeats to infinity its absolute value is defined exactly by $0{\cdot}\dot{2}$. Similarly 1/3 can be portrayed as $0{\cdot}\dot{3}$, but note however that 2/7 is $0{\cdot}\dot{2}\dot{8}\dot{5}\,\dot{7}\dot{1}\dot{4}$, signifying the repetition of all six digits:

0·285 714 285 714 285 714 285 714 etc.

The concept of a decimal fraction repeating to infinity thus poses no great problem. Irrational numbers on the other hand cannot be represented by fractions no matter how long, nor can their decimal portrayal be complete since there is no repetition or recurrence. However, although such a number possesses an infinite and non-repeating number of digits its magnitude is *finite*. All the natural numbers less than infinity have finite magnitude. It is theoretically possible for example to portray such numbers on a scale of some description whilst it is impossible to conceive of, much less accurately to represent, a string of digits whose length is infinite. Problems arise however when we consider that a string comprising an infinite number of digits may have as a result a magnitude which exceeds infinity. In practice however, infinity provides a convenient concept which may be used in the development of mathematical theory. If the reader is interested in following up some of the paradoxes of the infinite he is referred to Wilder (1968).

1.5 Approximation and accuracy

Examples of irrational numbers are to be found in $\sqrt{2}$ or in π. We frequently *approximate* $\sqrt{2}$ by abbreviation to 1·414. This is $\sqrt{2}$ expressed to three significant decimal places or to a resolution of 10^{-3}. We can more precisely approach the true value of $\sqrt{2}$ as we extend the resolution to say 10^{-15}, in which case we obtain 1·414 213 562 369 326. Approximation is commonly used if for no other reason than to save space and the tedious writing of long digit strings! It is however important on occasions to specify the resolution to which data are given, and a more detailed treatment of precision and approximation may be found in the companion book to this volume, Davidson (1978). Any reduction in resolution will of necessity involve 'rounding' the decimal fractions, so that to approximate to say six decimal places we require to know the value of the digit in the seventh place. For values of the '$(n+1)$th' place greater than or equal to five, the value of the 'nth' place is increased by one. Thus, $\sqrt{2}$ to six significant figures is $1{\cdot}41421\underline{4}$ and not $1{\cdot}41421\underline{3}$, since the seventh place has the value five. Similarly an approximation of 2·98 to one place of decimals is 3·0.

1.6 Roots and logarithms

Within physical geography many power terms are met which involve real powers. For example, Dury (1959) has shown that for the Nene and Great Ouse catchments in East Anglia, the discharge with a 2·33 year return period was proportional to the drainage area in the following way:

$$q = 5{\cdot}1a^{0{\cdot}98} \tag{1.1}$$

where q is the discharge in cusecs (cubic feet per second) and a is the contributing drainage area in square miles. The original Imperial units are adhered to in order to avoid confusion. The flood with a return period (see chapter 7) of 2·33 years has been shown to equal the mean annual flood of a river, and is thus of great importance in hydrological forecasting. The concept of raising a number to the power 0·98 is a difficult one to grasp and is best established with reference to more simple decimal powers, like 0·5 or $0{\cdot}\dot{3}$. Raising a number to the power 0·5 is the same as taking its square root: $x^{0{\cdot}5} = \sqrt{x}$. Similarly $x^{0{\cdot}\dot{3}} = x^{1/3} = \sqrt[3]{x}$, $x^{0{\cdot}25}$ is the fourth root, $\sqrt[4]{x}$ and so on. Further examples may be of use here since very frequently the ability to portray the same power in different ways leads to some confusion. Let us take first of all the value of 2^3. Elementary calculation leads to the solution of eight $(2 \times 2 \times 2)$. Thus we may also write: $\sqrt[3]{8} = 8^{1/3} = 8^{0{\cdot}\dot{3}} = 2$. Now since $\sqrt{4} = 4^{1/2} = 2$, then $(4^{1/2})^3 = \left(\sqrt{4}\right)^3 = 4^{1{\cdot}5} = 8$. Similarly $3^3 = 9^{3/2} = 27$ and $4^5 = 16^{5/2} = 1{,}024$, and so on. We can now define one important law of computation involving powers in that for the general case:

$$(a^m)^n = a^{mn} \tag{1.2}$$

Two further laws affect computation with expressions containing powers. It can be shown arithmetically for example that 2^5 is the same as $2^2 \times 2^3$, or $(2 \times 2) \times (2 \times 2 \times 2)$. In general terms therefore we may write:

$$a^m \, a^n = a^{(m+n)} \tag{1.3}$$

and similarly:

$$a^m / a^n = a^{(m-n)} \tag{1.4}$$

So returning to the example of the Nene and Great Ouse catchments we can see that $a^{0\cdot98}$ can be *approximated* as:

$$\sqrt{a}\sqrt[3]{a}\sqrt[7]{a} \text{ or } a^{0\cdot5}a^{0\cdot\dot{3}}a^{0\cdot14} = a^{0\cdot97}$$

We now have a particularly useful aid to computation involving multiplication and division. If as in the example above, we can express numbers which must be multiplied or divided as powers of a constant base, then addition of these powers will permit the multiplication of the original numbers and subtraction of the powers will facilitate division. These power values expressed to a constant base are called *logarithms* of the original numbers. For example, the very elementary product of $10 \times 100 = 1{,}000$ can be expressed as $10^1 \times 10^2 = 10^3$, using the common base of ten. The power to which the base ten is raised is thus the logarithm of that number to this base. Thus the logarithm of 100 to this base is 2·0, or:

$$\log_{10}100 = 2\cdot0 \text{ as } 10^2 = 100$$

Similarly:

$$\begin{aligned} \log_{10}10 &= 1\cdot0 \text{ as } 10^1 = 10 \\ \log_{10}1 &= 0\cdot0 \text{ as } 10^0 = 1 \\ \log_{10}0\cdot1 &= -1\cdot0 \text{ as } 10^{-1} = 0\cdot1 \end{aligned}$$

and so on. The addition of logarithms of numbers to a constant base enables us to solve for the product, and their subtraction the division. In general terms therefore:

$$\begin{aligned} \text{if } x^m = a &\quad \text{then } \log_x a = m \\ x^n = b &\quad \text{then } \log_x b = n \end{aligned}$$

More formally, the *logarithm to the base x of a number is the power to which x must be raised to obtain that number.* Further, we may state that:

$$\log_x a + \log_x b = \log_x(ab) \tag{1.5}$$

and

$$\log_x a - \log_x b = \log_x(a/b) \tag{1.6}$$

The most common value used for x is ten since manipulation in and out of logarithms is easy for this base. Sometimes however the base e is used. The derivation of the value of e appears in chapter 6 and for the present it is sufficient to give its approximate value as 2·718 28 to five decimal places. Such logarithms are the *Napeirian* or *natural* logarithms and are generally written $\log_e x$ or $\ln x$. Generally though if the base is not specified the base ten is assumed, and we shall now confine our attention to these base ten logarithms.

It is a relatively simple conceptual step to consider now the logarithms for numbers other than those which are exact multiples of the base, although the powers involved are at first sight rather alarming decimals. Clearly, for example, the logarithms of all numbers between one and ten must lie between 0·0 and 1·0, those for numbers between ten and 100 extend from 1·0 to 2·0 and so on, so that the overall realm for each is quite clearly defined already. The logarithms between these values in each group can be calculated, but they are more usefully and conveniently found in table form, generally to four or more decimal places (e.g. 'four-figure logarithm tables' are significant to four decimal places and are given in Appendix 3). A small portion of the tables is reproduced in table 1.2 to illustrate the technique of obtaining the logarithm of a number. We can use this particular portion to find the logarithm of any number from 5·200 to 5·499. The pairs of digits in the extreme left-hand column represent the first two digits of the number for which we require the

Table 1.2 Extract from four-figure logarithm tables.

	0	1	2	3	4	5	6	7	8	9	1	2	3	4	5	6	7	8	9
52	0·7160	0·7168	0·7177	0·7185	0·7193	0·7202	0·7210	0·7218	0·7226	0·7235	1	2	2	3	4	5	6	7	7
53	0·7243	0·7251	0·7259	0·7267	0·7275	0·7284	0·7292	0·7300	0·7308	0·7316	1	2	2	3	4	5	6	6	7
54	0·7324	0·7332	0·7340	0·7348	0·7356	0·7364	0·7372	0·7380	0·7388	0·7396	1	2	2	3	4	5	6	6	7

logarithm. The next group of columns, headed 0 to 9, portray the magnitude of the third digit of the number, and the final group, the magnitude of the final digit. The numbers beneath the first group of columns represent the four-figure logarithms of numbers up to three digits in length. The decimal point is omitted but would occur to the left of the number. Thus log 5·30 can be obtained by reading the row commencing with 53, under column 0, and is 0·724 3. Similarly log 5·36 is 0·729 2. For numbers containing strings of four digits the values in the relevant column of the second group must be *added* to the previous three-digit score. Thus log 5·301 is 0·7243 plus 0·0001 ($=0{\cdot}7244$), and $\log 5{\cdot}365 = 0{\cdot}7292+0{\cdot}0004 = 0{\cdot}7296$. Reference to the tables in Appendix 3 will now reveal that:

$$\begin{aligned}\log 2 &= 0{\cdot}301\,0 \text{ or } 10^{0\cdot 301\,0} = 2\\ \log 3 &= 0{\cdot}477\,1 \text{ or } 10^{0\cdot 477\,1} = 3\\ \log 5 &= 0{\cdot}699\,0 \text{ or } 10^{0\cdot 699\,0} = 5\\ \log 5{\cdot}1 &= 0{\cdot}707\,6 \text{ or } 10^{0\cdot 707\,6} = 5{\cdot}1\end{aligned}$$

Similarly the numbers between 10·0 and 99·9 all lie between 1·0 and $1{\cdot}\dot{9}$. Here we must bear in mind the additional factor of ten, or logarithm of 1·0 so that for example, $\log 22{\cdot}0 = 1{\cdot}342\,4$, since $22{\cdot}0 = 2{\cdot}20 \times 10^1$, and therefore, $\log 22{\cdot}0 = \log 2{\cdot}20 + \log 10^1 = 0{\cdot}342\,4 + 1{\cdot}000\,0$. Similarly the logarithms of numbers less than one may be found as follows:

$$\begin{aligned}\log 0{\cdot}036\,4 &= \log 3{\cdot}64 + \log 10^{-2}\\ &= 0{\cdot}561\,1 + (-2{\cdot}000\,0)\end{aligned}$$

which is written: $\bar{2}{\cdot}561\,1$ ('bar two point 561 1'). Note here in particular that the right-hand side of the logarithm is still *positive* even though $\bar{2}$ indicates that the left-hand side is not. Thus to solve $0{\cdot}012\,3 \times 0{\cdot}001\,9/0{\cdot}240\,1$ by logarithms we have (by reference to appendix 3),

$$\begin{aligned}\log 0{\cdot}012\,3 &= -2+0{\cdot}089\,9 = \bar{2}{\cdot}089\,9\\ \log 0{\cdot}001\,9 &= -3+0{\cdot}278\,8 = \bar{3}{\cdot}278\,8\\ \log 0{\cdot}240\,1 &= -1+0{\cdot}380\,4 = \bar{1}{\cdot}380\,4\end{aligned}$$

so that:

$$\begin{aligned}\log(0{\cdot}001\,9/0{\cdot}240\,1) &= (\bar{3}{\cdot}278\,8)-(\bar{1}{\cdot}380\,4)\\ &= (-3+0{\cdot}278\,8)-(-1+0{\cdot}380\,4)\\ &= -2+(0{\cdot}278\,8-0{\cdot}380\,4)\\ &= -2-1+0{\cdot}898\,4 = \bar{3}{\cdot}898\,4\end{aligned}$$

and $\log(0{\cdot}012\,3 \times 0{\cdot}001\,9/0{\cdot}240\,1)$ is therefore $\bar{2}{\cdot}089\,9 + \bar{3}{\cdot}898\,4$. Adding the left-hand side and right-hand sides separately we have:

$$\bar{2}{\cdot}089\,9 + \bar{3}{\cdot}898\,4 = -5+0{\cdot}988\,3 = \bar{5}{\cdot}988\,3$$

Reversing the process now, and taking the *antilogarithm* from tables we have:

$$\text{antilog}\,(\bar{5}{\cdot}988\,3) = 9{\cdot}734 \times 10^{-5}$$

If logarithms are used in ways other than as an aid to computation as above then the mixture of signs on either side of the decimal point must be taken into account. In a later chapter we shall see that the solution of equations involving a logarithm term demands that the logarithms be converted into a real number, so that for example $\log(0{\cdot}011) = \bar{2}{\cdot}041\,4$ which becomes $0{\cdot}041\,4 - 2{\cdot}000\,0 = -1{\cdot}958\,6$.

A further use of logarithms is to enable us to calculate the value of real roots and powers of numbers. Consider equation (1.2) and let:

$$(x^m)^n = x^{mn}$$

Therefore

$$\text{if } a = x^m, \log_x(a^n) = n \log_x a \tag{1.7}$$

So, for example, $2^{0\cdot5}$ can be found by multiplying log 2 by 0·5:

$$\log 2 = 0{\cdot}3010 \text{ (from tables)}$$
$$\therefore\ 0{\cdot}5 \log 2 = 0{\cdot}1505$$

and

$$2^{0\cdot5} = 1{\cdot}415 \quad \text{(from tables)}$$

Similarly for $54{\cdot}6^7$:

$$\log(54{\cdot}6^7) = 7 \log 54{\cdot}6$$
$$= 7 \times 1{\cdot}7372$$
$$= 12{\cdot}2$$
$$\text{and } \therefore\ 54{\cdot}6^7 = \text{antilog } 12{\cdot}2 = 1{\cdot}445 \times 10^{12}$$

We are now in a position to solve equation (1.1) for a known drainage area, say when $a = 22{\cdot}0$ square miles. Taking the logarithm of both sides of the equation we have:

$$\log q = \log 5{\cdot}1 + 0{\cdot}98 \log 22{\cdot}0$$
$$= 0{\cdot}7076 + 0{\cdot}98 \times 1{\cdot}3424$$
$$= 2{\cdot}0232$$
$$\text{and } \therefore\ q = 105{\cdot}4 \text{ cusecs (from tables)}$$

The reader may now wish to establish for himself using the tables in appendix 3, the following logarithms, and in the case of numbers less than one, their real equivalent, and then practice using logarithms to solve arithmetic problems.

Problems 1.6

1.6.1 log 0·0150
1.6.2 log 12·24
1.6.3 log 56·847
1.6.4 log 0·0025
1.6.5 log $(5{\cdot}4 \times 10^{-2})$
1.6.6 log (54×10^{-3})

Now work through the following problems.

1.6.7 Find the solution of $2{\cdot}56 \times 3{\cdot}012/10{\cdot}8$ using logarithms.
1.6.8 Solve $0{\cdot}8024 \times 0{\cdot}0543$ using logarithms.
1.6.9 Determine $\sqrt{2}$, $42{\cdot}36^3$, $5{\cdot}25^4$ and $\sqrt[3]{3}$ using logarithms.
1.6.10 Using logarithms show that in equation (1.1) when $a = 300$ square miles, $q = {\cdot}1{,}365$ cusecs.

1.7 Elementary functions

The expression used in equation (1.1) was introduced to illustrate the nature of powers and logarithms of numbers. Equation (1.1) also expresses a relationship between a, contributory drainage area, and q, river discharge. We are able to solve this equation for q for any value of a which lies within the range of values used by Dury to determine the relationship. q and a are *variables* and indicate that for any *one* value of a, q has an equivalent value defined by the equation. Each can adopt any one of a number of values between the limits used to establish the relationship (the largest and smallest drainage basins). Thus we say the a always has a value lying within a group of values called *set* A and q a value within the group of values comprising *set* Q. Thus, a is a member of A and q is a member of Q, or:

$$a \in A$$
$$q \in Q$$

and within these constraints:

$$q = 5{\cdot}1a^{0{\cdot}98}$$

This equation is a *function* relating q to a or:

$$q = f(a)$$

In effect we are saying that q is proportional to a ($q \propto a$) or q varies as a.

Very often in physical geography more than one factor is seen to influence variation in another. Frequently therefore we find functions which are composed of more than one variable. For example, we may argue that soil type (s) is influenced by climate (c), vegetation (v), organisms (o), parent material (p), rainfall (r) and time (t). Without saying in what way each of these influences soil type we can abbreviate this to:

$$s = f(c,v,o,p,r,t)$$

We would then attempt to set up field experiments to examine in which way soil type is influenced by each of these—a good starting point for a research design. Any mathematical function in equation form, such as (1.1) specifies that a definite relationship exists between one or more variables. For the pure mathematician this certainty is important, but as physical geographers we must be aware that there are almost always indefinable and unknown factors entering most relationships with which we deal. This raises a number of important points which must lead us to doubt the geographical certainty of many mathematical functions which purport to represent a geographical relationship. Firstly, we know from geographical reasoning that whilst q depends upon a, the reverse cannot be said to be true. Thus the mathematical function must take on more the meaning of a logical one-way 'implication'. An implication is the term given to a statement which says that 'if there is an increase in a, then there will be a corresponding increase in q'. There are ways of symbolizing such expressions and if the reader particularly wishes to follow up this aspect of logic he is referred to one of the suggested texts on the subject under the heading of further reading at the end of the book. We generally call the variable which induces the change in the other or others, the *independent* variable, and the one in which change is induced, the *dependent* variable.

Secondly, the equation which Dury derived, and others like it, was a result of a relatively small sample taken from a much larger population of drainage basins. There is thus again no absolute certainty that the relationship is incontrovertible as the mathematical function describing it suggests. This as we shall see in chapter 3, means we must be guarded in interpreting such expressions. Thirdly, there is a number of other factors which can also influence the magnitude of the mean annual flood, for example, precipitation and its seasonal distribution and type. This throws a further emphasis on the one-way nature of the relationship. Large drainage basins may well be associated in general with a mean annual flood of great magnitude, but merely to observe that a certain basin has a very large mean annual flood does not necessarily indicate a large contributory drainage area.

Returning now to the form of the function itself, simply to say that $q = f(a)$ or that $y = f(x)$ merely states that some sort of relationship exists. It does not specify the relationship. For example, it does not state:

(i) by how much y varies for unit change in x (or q with a);
(ii) whether an incremental change in x always produces the same incremental change in y, and if not, then the trend in incremental changes in y with x; and
(iii) whether when x is zero, y is zero.

Equation (1.1) was specific, it stated that:

$$q = f(a^n), \text{ where } n \neq 0 \text{ or } 1 \qquad (1.8)$$

If n had been equal to one instead of 0·98, then in the relationship:

$$q = 5{\cdot}1a$$

unit change in a would produce corresponding changes in q which were always 5·1 times a (see table 1.3). This type of function is called a *linear* function for reasons which will become clear in chapter 3. Note however, that when $a = 0$, $q = 0$. If the equation had been, $q = 5{\cdot}1a + 6{\cdot}4$, then when $a = 0$, $q = 6{\cdot}4$: $a = 1$, $q = 5{\cdot}1 + 6{\cdot}4 = 10{\cdot}5$ and so on. This is the addition of a *constant* term in the equation and gives us a general form for a linear function of:

$$y = mx + k,$$

which we shall follow up in chapter 3. The term m in the equation satisfies condition (i) above, and the constant term k satisfies condition (iii). Condition (ii) does not apply in this case, since we have established that incremental changes are constant for all values of x.

Table 1.3 Table of discharge (cumecs) for a given drainage area (km).

a	1	2	3	4	5	6
q	5·1	10·2	15·3	20·4	25·5	30·6

We know however, from equation (1.8) that $n \neq 0$ or 1 and thus that q is a *power function* of a. This means that unit changes in a do not produce constant incremental changes in q. Again for reasons which will soon become evident a power function is a *curvilinear* function, or a *standard* function. These types of function are developed fully in chapter 5. Where the value of n is greater than zero we have a situation of *direct proportionality*, that is to say *increases* in a always induce *increases* in q. If $n = 0{\cdot}0$ then q would be constant, since any value raised to the power zero, is unity. If however n is less than zero inverse proportionality holds. An *increase* in a produces a *decrease* in q. We now have two types of curvilinear functions which at their most elementary are:

$$y = x^n \text{ (direct proportionality, } n>0)$$
$$y = x^{-n} \text{ (inverse proportionality)}$$

Many derived relationships in the environmental sciences are curvilinear, and there are numerous other types of such functions, such as:

$$y = \log x$$
$$y = e^x$$
$$y = \sin x$$

The actual form of these is followed up in chapter 5.

1.8 Coordinates and graphs

Any mathematical function can be portrayed graphically with reference to *axes*, and indeed much of the terminology used in describing mathematical

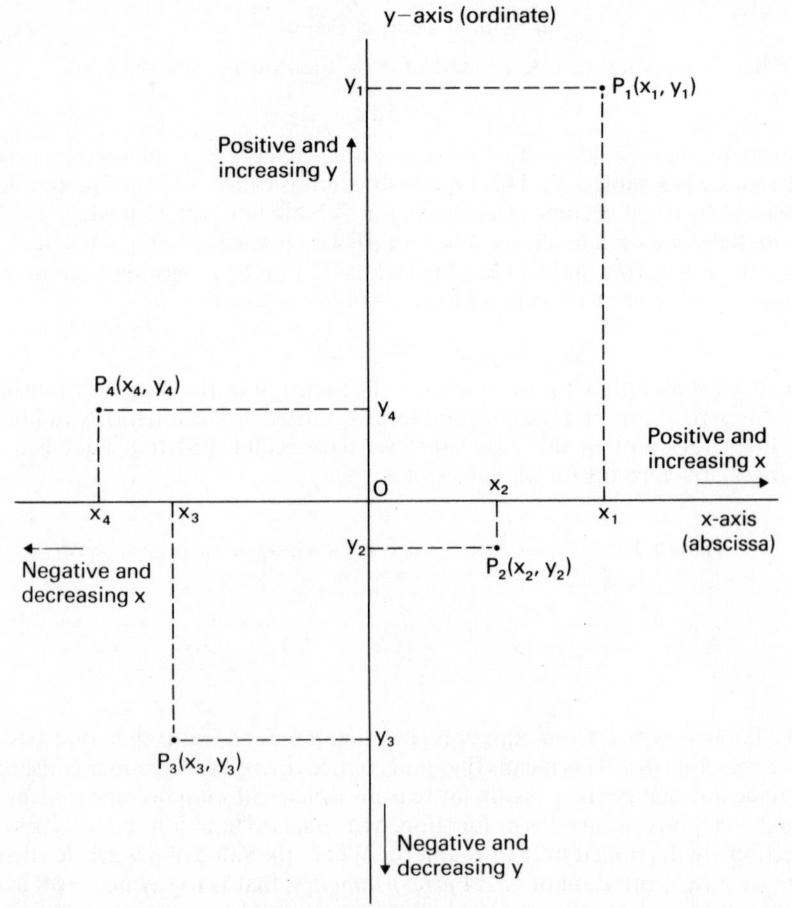

Figure 1.2 Rectangular coordinate grid.

functions in section 1.7 relied heavily upon the visual impact of functions when plotted on graphs. Hence, in the previous section, we referred to linear functions, which plot as a straight line, and curvilinear functions, which adopt a curved form on a graph. Graph axes provide reference points in the same way that eastings and northings do for topographic maps, except that grid references are replaced by *coordinates*. The magnitudes of variables are plotted along the axes of the graph, such that one variable is portrayed on each axis. There will be as many axes as there are variables in the function to be plotted, and although many relationships in physical geography are *multivariate* in that they involve more than two variables, we shall for the time being at least, consider only *bivariate* relationships, as these are less confusing to show on the two-dimensional pages of a book.

The most commonly used system utilizes the *rectangular* or *Cartesian* grid, whose axes are at right-angles to each other. The vertical axis, the y-axis, or ordinate, always represents the dependent variable, whilst the horizontal axis, the x-axis or *abscissa*, always represents the independent variable. Axes intersect at a point known as the *origin* where values on both axes are zero. The y-axis below the origin is always negative, and the x-axis to the left of the origin is always negative. Thus in figure 1.2 point P_1 has coordinates (x_1,y_1) which are both positive, whilst P_3 has coordinates (x_3,y_3) which are both negative. In the remaining two quadrants P_2 (x_2,y_2) has a negative y-coordinate and a positive x-coordinate, and P_4 (x_4,y_4) has a positive y-coordinate and a negative x-coordinate. The visual form of any function can be portrayed with reference to these axes by plotting successive coordinate pairs.

A second and far less commonly used system is that which uses *polar coordinates*. Cartesian coordinates offer a far more flexible system. In the polar

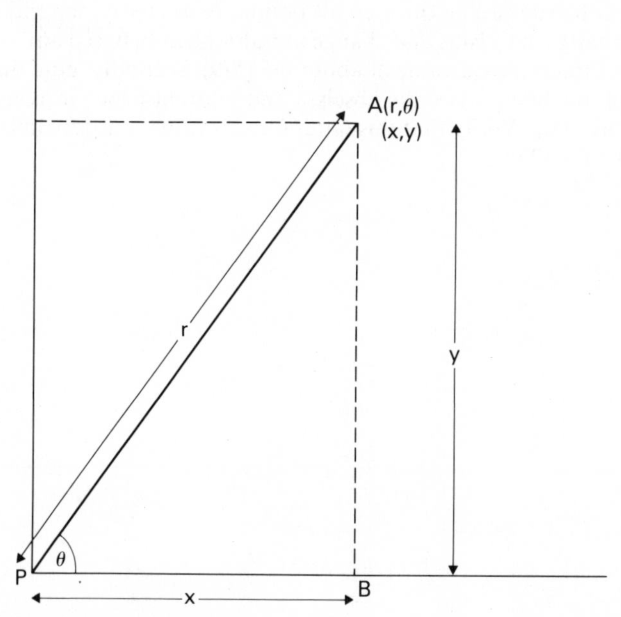

Figure 1.3 Polar coordinate grid.

graph in figure 1.3, point A is defined by the coordinates (r, θ) where: r is the distance from the *pole* (P) to A, and θ is the *vectorial angle* subtended by the line AP to the *initial line* PB. Circular graphs using a polar grid are used in certain aspects of physical geography for example, wind roses and stone orientation diagrams. Confusion can arise however between conflicting conventions for measuring orientation in degrees *clockwise* from the *north* point (0 degrees), such that south equals 180 degrees, west 270 degrees and so on, and the *anticlockwise* measurement from the horizontal initial line shown in figure 1.3. The main use of the polar grid system is found in expressing *vector* quantities (chapter 2), for example in resolving wind direction and velocity.

Problems 1.8

The reader may now wish to plot these two simple examples of linear functions.

1.8.1 Construct x- and y-axes between zero and six, and zero and 40 respectively, and plotting q on the y-axis and a on the x-axis, attempt a reconstruction of $q = 5.1a$, whose values between these limits appear in table 1.3. Verify graphically that when a equals about four, q approximately equals 20.

Table 1.4 Equivalent Celsius and Fahrenheit values.

°C	−9·4	−3·9	1·7	7·2	12·8	18·3	23·9	29·4	35·0
°F	15	25	35	45	55	65	75	85	95

1.8.2 The data in table 1.4 show Fahrenheit temperatures and their Celsius equivalents. Plot Celsius on the ordinate and Fahrenheit on the abscissa, and construct a line through all points. Note firstly, the ratio between the change in Celsius and change in Fahrenheit between any two points. This should have a value of about $0{\cdot}\dot{5}$ (5/9). Secondly, note the point at which the line crosses the abscissa and point at which it intersects with the ordinate. Verify that this latter Celsius value is approximately equal to $0{\cdot}\dot{5}$ $(-32{\cdot}0)$.

2 Trigonometry and vectors

Movement and spatial measurement are intrinsically linked with studies in physical geography. Studies of process involve movement, whilst the nature of geography itself inevitably involves a consideration of process and form in the spatial context. The measurements of length, area and volume are of course convenient means of providing spatial measurement, and as we saw at the end of chapter 1 we may pinpoint any location in space with reference to coordinates on a graph. However, we also ended the previous chapter by establishing that points may be equally precisely defined with reference to a line of length r subtending an angle θ to a similar line whose alignment is already established. This means of representation can be developed into *vector* quantities. In their turn vectors lead us on to a study of *matrices* in chapter 4. Both vectors and matrices add, quite literally, a further dimension to the mathematics so far developed in this book, and provide for a better understanding of mathematical principles by giving an alternative viewpoint.

The mathematical range contained within each of the topics in this chapter is considerable. There are for example, many specifically mathematical texts at present on the market which deal entirely with vectors and matrices: Knight and Adams (1975) is for example, a relatively simple text. All we are able to do in this book is to develop the elementary mathematics of both vectors and matrices within the confines of two chapters, this one and chapter 4. An understanding of vectors as applied to the dynamics of the physical environment considerably helps in the interpretation of the processes involved, whether they be geomorphological, hydrological or from any other aspect of physical geography. The development of the principles behind matrix algebra will enable the reader to attempt a deeper understanding of multivariate statistical methods, such as principal components analysis, and will additionally provide a useful alternative viewpoint of multivariate 'space'. To those interested in the use of computers for the analysis of data a knowledge of matrices will also help them to understand some aspects of computer programming. All we must hope to attain here is the establishment of the basic principles of vectors and matrices. It is left to the reader to develop the topics to a more advanced level through the further reading of papers in geographical journals and of mathematical and statistical texts. To this end some suggestions for further reading are given at the end of chapter 4. First of all however we must define a number of angular measures and trigonometric ratios.

2.1 Angular measurement; degrees and radians

For much elementary work involving the trigonometric ratios it is sufficient to measure angles in degrees, minutes and seconds, for example, 24° 34′ 02″, where 60 seconds (″) equal one minute (′) and 60 minutes equal one degree. 360 degrees describe a complete circle. However at more advanced levels and especially when the trigonometric ratios are treated as functions, as for example in $y = \sin x$, angles are measured in *radians*. A radian is defined as the angle subtended at the centre of a circle which contains a section of the circumference of the circle of length equal to the radius, r (see figure 2.1). Thus, since the circumference of the circle is $2\pi r$, the complete circle of 360 degrees is given by: $2\pi r/r$ radians, or, $360° = 2\pi$ radians. So, $90° = \pi/2$ radians and $180° = \pi$ radians, and one degree is 0·0175 radian or one radian is 57° 17′ 45″ = 57·296°.

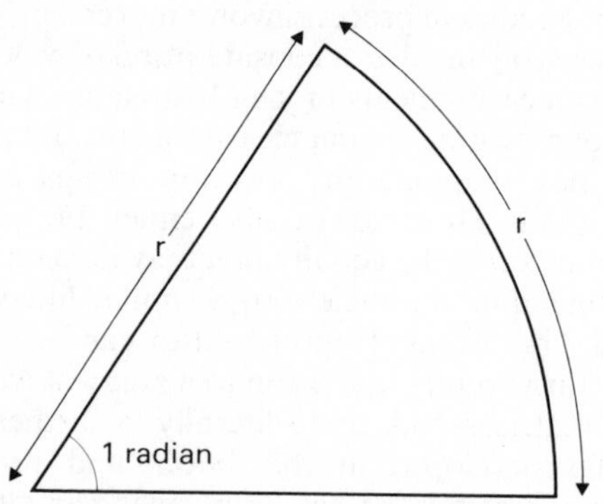

Figure 2.1 Definition of a radian.

2.2 Trigonometric ratios

Certain basic trigonometric ratios require to be defined before a full treatment of vectors is possible and so that mathematical functions may be considered in later chapters. The vectorial angle θ in figure 1.3 serves to illustrate these basic ratios. This also permits us to determine an expression for rectangular coordinates (x,y) from the polar coordinates (r,θ). As the value of θ changes, so too do the ratios between the three sides of the triangle PAB; PA, AB, and BP (respectively the hypoteneuse, opposite and adjacent sides to the angle APB). The angle APB is a right-angle. As θ approaches zero degrees side AB becomes very small, whilst sides PA and BP are almost equal, although by definition length BP is always less than PA. When θ is 45°, the lengths of AB and BP are the same and the triangle becomes *isosceles*. Thus comparison of the lengths of any pair of sides in the form of a trigonometric ratio permits the determination of the magnitude of θ by the use of tables. The reader may determine the values of the basic three trigonometric ratios (sine, cosine and tangent) for angles between zero and 90 degrees by using the tables in appendix 4. The method of using these is similar to that described in section 1.6 for the derivation of logarithms. The ratios of sides of the triangle PAB for each of the basic trigonometric ratios are:

$$\text{sine or } \sin\theta = AB/AP = y/r \tag{2.1}$$

$$\text{cosine or } \cos\theta = BP/AP = x/r \tag{2.2}$$

$$\text{tangent or } \tan\theta = \text{AB/BP} = y/x \tag{2.3}$$
$$\text{cosecant or } \operatorname{cosec}\theta = \text{AP/AB} = r/y \tag{2.4}$$
$$\text{secant or } \sec\theta = \text{AP/BP} = r/x \tag{2.5}$$
$$\text{cotangent or } \cot\theta = \text{BP/AB} = x/y \tag{2.6}$$

Since, $\cos\theta = x/r$

then, $x = r\cos\theta$ (multiplying by r)

and also, $y = r\sin\theta$

so that for a rectangular grid, point A with polar coordinates (r,θ) has rectangular coordinates $(r \cos\theta, r \sin\theta)$ assuming point P corresponds precisely to the Cartesian origin O. Those who have difficulty in remembering the above list of basic trigonometric ratios would do well to remember firstly that $\operatorname{cosec}\theta = 1/\sin\theta$, $\sec\theta = 1/\cos\theta$, and $\cot\theta = 1/\tan\theta$, and secondly commit to memory the mnemonic, SOHCAHTOA: Sine, Opposite/Hypoteneuse; Cosine, Adjacent/Hypoteneuse; Tangent, Opposite/Adjacent.

Using the same framework, Pythagoras' theorem states that:

$$r^2 = x^2 + y^2 \tag{2.7}$$

or, the square of the hypoteneuse is equal to the sum of the squares of the other two sides, and therefore that (dividing through by r^2):

$$1 = \frac{x^2}{r^2} + \frac{y^2}{r^2}$$

and thus (from equations (2.1) and (2.2)):

$$\cos^2\theta + \sin^2\theta = 1 \tag{2.8}$$

or: $$\cos^2\theta = 1 - \sin^2\theta$$

and: $$\sin^2\theta = 1 - \cos^2\theta$$

Also, since $$\tan\theta = \frac{y}{x}$$

$$\therefore \tan\theta = \frac{y/r}{x/r}$$

and thus: $$\tan\theta = \frac{\sin\theta}{\cos\theta} \tag{2.9}$$

From equation (2.8), and dividing by $\cos^2\theta$ we obtain:

$$\frac{\sin^2\theta}{\cos^2\theta} + \frac{\cos^2\theta}{\cos^2\theta} = \frac{1}{\cos^2\theta}$$

$$\therefore \tan^2\theta + 1 = \sec^2\theta \tag{2.10}$$

and also $$\therefore \cot^2\theta + 1 = \operatorname{cosec}^2\theta \tag{2.11}$$

The above relationships will be familiar to many readers already and are important in later chapters, and if not already known should be committed to memory. Relationships can also be established between combinations of angles. Consider the case in figure 2.2(a) with the compound angle $(\alpha + \beta)$. It can be shown that:

$$\sin(\alpha+\beta) = \sin\alpha\cos\beta + \cos\alpha\sin\beta \tag{2.12}$$

$$\cos(\alpha+\beta) = \cos\alpha\cos\beta - \sin\alpha\sin\beta \tag{2.13}$$

and

$$\tan(\alpha+\beta) = \frac{\tan\alpha+\tan\beta}{1-\tan\alpha\tan\beta} \tag{2.14}$$

Now, for figure 2.2(b), where angle α includes angle β, then:

$$\sin(\alpha-\beta) = \sin\alpha\cos\beta - \cos\alpha\sin\beta \tag{2.15}$$

$$\cos(\alpha-\beta) = \cos\alpha\cos\beta + \sin\alpha\sin\beta \tag{2.16}$$

and

$$\tan(\alpha-\beta) = \frac{\tan\alpha-\tan\beta}{1+\tan\alpha\tan\beta} \tag{2.17}$$

Further, if angle α = angle β then:

from (2.12) $$\sin 2\alpha = 2\sin\alpha\cos\alpha \tag{2.18}$$

from (2.13) $$\cos 2\alpha = \cos^2\alpha - \sin^2\alpha$$

$$\cos 2\alpha = 1-2\sin^2\alpha = 2\cos^2\alpha - 1 \tag{2.19}$$

and from (2.14) $$\tan 2\alpha = \frac{2\tan\alpha}{1-\tan^2\alpha} \tag{2.20}$$

For the proofs of these theorems the reader is referred to Flanders and Price (1973) in the list of books for further reading.

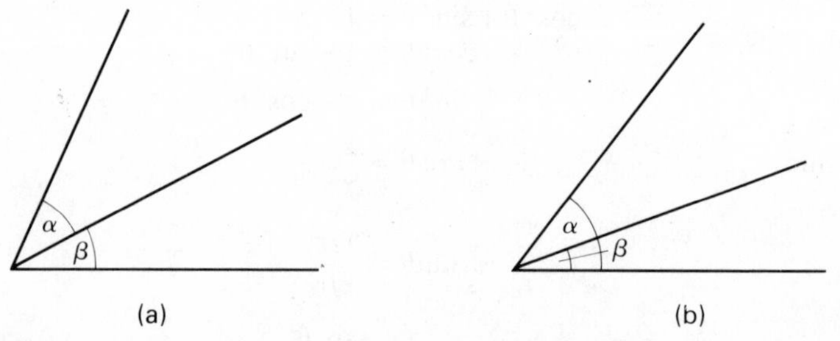

Figure 2.2 Trigonometric ratios of compound angles: (a) $(\alpha+\beta)$; (b) $(\alpha-\beta)$.

The trigonometric ratios above have been derived in terms of angles of less than 90 degrees. Clearly however it is possible to take any trigonometric ratio of any angle between zero and 360 degrees (0 to 2π radians). Such values can be derived using equations (2.7) to (2.20) by substituting for α, 90, 180, or 270 in the form sin 146 = sin(90+56) and so on. The detailed treatment of this is better illustrated using trigonometric ratios as trigonometric functions of the form $y = \sin x$, $y = \cos x$ and $y = \tan x$. Such functions are *periodic* in that they have a regular cyclical recurrence, in a similar way to repeating decimals. For example the 'sine curve' of $y = \sin x$ is shown in figure 2.3. From this it can be seen that more than one angle has the sine 1·0. In fact if we were to take x up to n radians, the angles for which the sine is 1·0 are $\pi/2$, $5\pi/2$, $9\pi/2$, etc. (or 90,

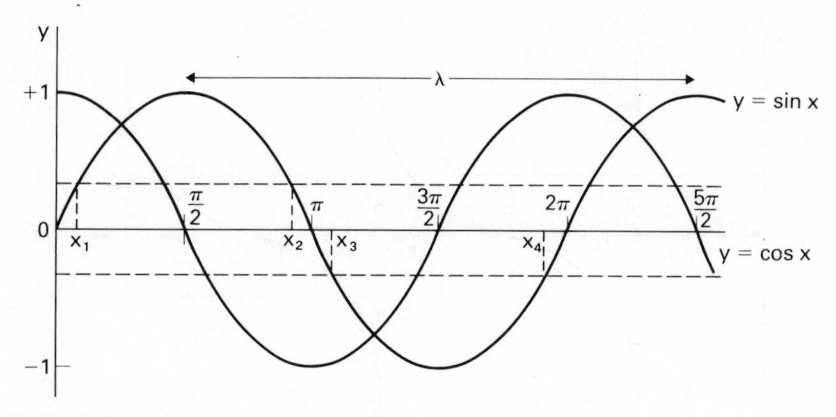

Figure 2.3 Graphs of the functions $y = \sin x$ and $y = \cos x$.

450, 810 degrees etc.). Further the angle which has sine value zero repeats every π radians at 0, π, 2π, 3π, etc., or every 180 degrees. A similar sequence exists for $y = \cos x$, but with the *phase* shifted by $\pi/2$ radians on the x-axis. The distance along the x-axis between adjacent troughs or adjacent ridges is termed the *wavelength* (generally represented by the Greek letter, λ) and in each case is 2π radians. The *amplitude* of the waveform is likewise the same in each case and equals two, between $y = +1$ and $y = -1$.

From the diagram it is equally clear that for any given value of y between 0 and 2π radians there exist always *two* angles (x_1 and x_2) for which y is the sine. A similar case arises where the sine value is negative (x_3 and x_4). We can thus see from the graph in figure 2.3 that the following are true:

$$\sin x_2 = \sin(180 - x_2) \tag{2.21}$$

$$\sin x_3 = -\sin(x_3 - 180) \tag{2.22}$$

$$\sin x_4 = -\sin(360 - x_4) \tag{2.23}$$

Similar relationships exist for cosine and tangent. The graph for $y = \tan x$ is shown in figure 2.4. These are:

$$\cos x_2 = -\cos(180 - x_2) \tag{2.24}$$

$$\cos x_3 = -\cos(x_3 - 180) \tag{2.25}$$

$$\cos x_4 = \cos(360 - x_4) \tag{2.26}$$

$$\tan x_2 = -\tan(180 - x_2) \tag{2.27}$$

$$\tan x_3 = \tan(x_3 - 180) \tag{2.28}$$

$$\tan x_4 = -\tan(360 - x_4) \tag{2.29}$$

Note that in the case of tangents the curve approaches *positive* infinity at $x = \pi/2$ and $3\pi/2$, and approaches from *negative* infinity immediately after these values. All three basic trigonometric ratios are positive between zero and $\pi/2$ radians, whilst each in turn is positive in the other three quadrants. The rules for determining trigonometric ratios for angles between 90 and 360 degrees are

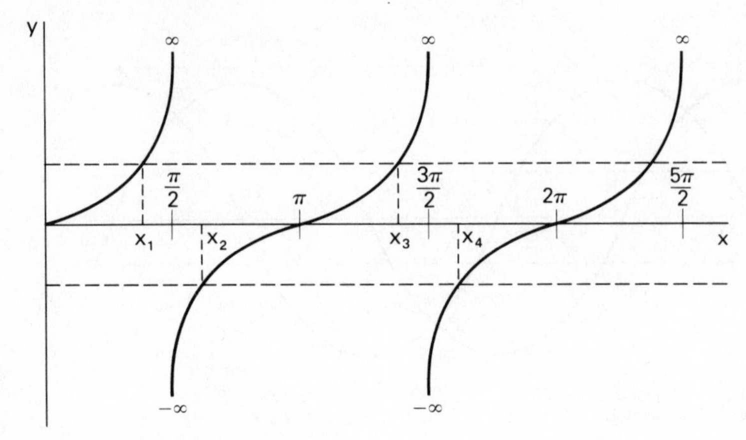

Figure 2.4 Graph of the function $y = \tan x$.

consistent for all ratios with the reservation that the sign of the ratio changes for different ratios in different quadrants. These are summarized in table 2.1 and figure 2.5. As a rule, to find say a trigonometric ratio of 121 degrees, find first the relevant ratio of $180 - 121 = 59$ degrees, and then note that only sine is positive in this quadrant. From tables we can thus see that sin 121 = sin 59 = 0·8572, but that tan 121 = −1·6643 and cos 121 = −0·5150. The reader is encouraged to familiarize himself with the trigonometric tables in appendix 4 before continuing further.

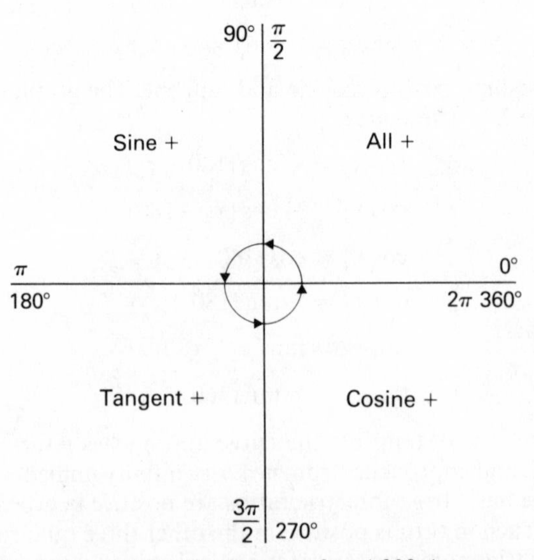

Figure 2.5 Positive trigonometric ratios between 0 and 360 degrees.

Table 2.1 Derivation of values of sine, cosine and tangent for all angles 0 to 360 degrees.

θ between	0–90	91–180	181–270	271–360
sine	$\sin\theta$	$\sin(180-\theta)$	$-\sin(\theta-180)$	$-\sin(360-\theta)$
cosine	$\cos\theta$	$-\cos(180-\theta)$	$-\cos(\theta-180)$	$\cos(360-\theta)$
tangent	$\tan\theta$	$-\tan(180-\theta)$	$\tan(\theta-180)$	$-\tan(360-\theta)$

2.3 Vectors and scalars

Thus far in the book we have dealt with numbers used simply to express quantity: for example, 85 cumecs, 100 kg or 25°C. These are all examples of *scalar* numbers merely signifying a certain specified *magnitude.* There also exists however, a further group, which, as well as indicating magnitude, also specify *direction.* These are *vectors* and have particular relevance to the climatologist when considering wind motion, or to the fluvial geomorphologist when he is assessing the effect of thalweg slope upon the erosive power of river discharge. To take the climatological example as a simple illustration, the statement that the wind is blowing at 10 m s^{-1} is to ascribe a scalar quantity to wind *speed*, but the statement that the wind has a *velocity* of 10 m s^{-1} from a northwesterly direction is a vector quantity. Note here the important difference between the uses of the words speed and velocity. A recent paper by Hindi and Kelway (1977) has used vector mathematics to attempt to correct or 'smoothe' the isochrones of rainfall resulting from the passage of a depression over England and Wales. These isochrones were derived from rather a sparse network of autographic raingauges whose time resolution is often poor. At the much broader scale many aspects of physical geography embrace and overlap with the elementary principles of mechanics or dynamics, particularly so in process studies. Forces (see Davidson 1978, chapter 3) are vector quantities and can be manipulated according to certain elementary principles. Much of the theory behind process studies in slope stability is bound up with the treatment of downslope forces under the influence of the acceleration due to gravity (g). We are perhaps used to seeing in such literature the expression $g \sin \theta$, where θ is the angle of slope the ground makes with the horizontal (figure 2.6). Similarly in meteorology one of the more commonly seen expressions (and often one of the least well understood by students!) is that used to

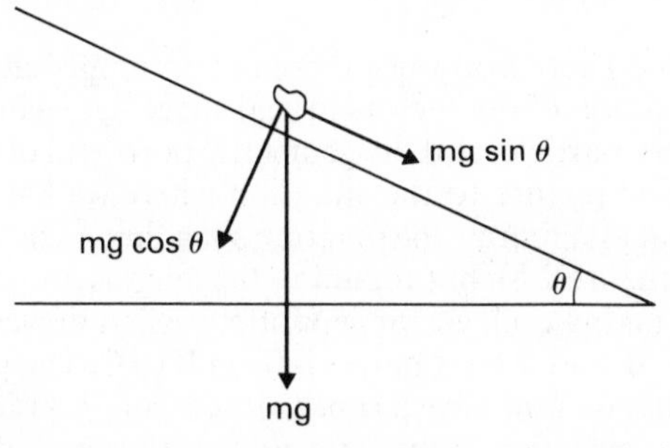

Figure 2.6 Forces acting on a particle on a slope, where *mg* is the weight of the particle and θ is the angle of slope.

describe in mathematical terms the variation of the Coriolis force over the earth's surface, $\omega \sin\phi$, where ω is the angular velocity of the earth and ϕ is the angle of latitude at the point at which the Coriolis force is being calculated. In this example the spin of the earth beneath a horizontally moving air particle imparts an apparent deflection to the path taken by the particle. However, the horizontal plane through which the particle is moving is *parallel* to the earth's axis at latitude 0, but at *right-angles* to it at the poles. The effect of the earth's rotation is therefore nil at the equator but at a maximum at the poles, where it equals the angular velocity of the earth. Some factor is therefore required which is proportional to latitude and varies between zero and one, and for reasons which we shall soon see, the sine of the latitude provides this factor. What we have done in defining the Coriolis parameter is to *resolve* the effect of the earth's rotation into the vertical plane. In this case the earth's angular velocity provides us with the vector quantity.

2.4 Vector algebra

In algebraic terms a vector quantity is always shown in bold (Clarendon) type. For example in figure 2.7, the vector represented by the line AB of magnitude v is depicted by $\mathbf{v}$ or $\overrightarrow{AB}$, or:

$$\mathbf{v} = \overrightarrow{AB} \tag{2.30}$$

indicating that the direction of the vector of magnitude v is from A to B. Similarly in the figure we have $\mathbf{w} = \overrightarrow{AC}$ and $\mathbf{z} = \overrightarrow{AD}$. We have portrayed each of these vectors in the figure such that the magnitude of each is indicated by the *length* of the lines depicting their direction. Thus $v = AB$, $w = AC$ and $z = AD$

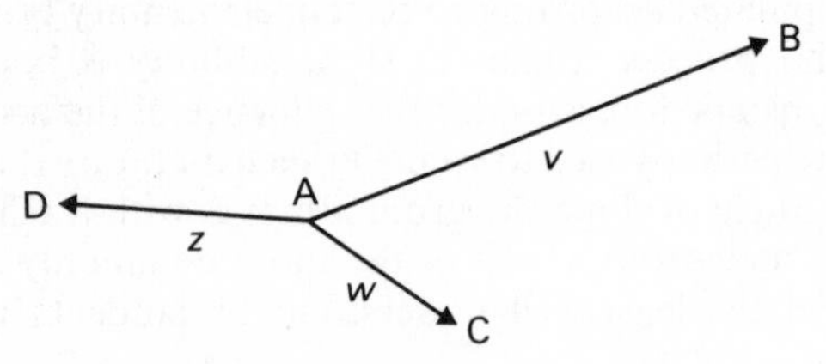

Figure 2.7 Simple vector diagram.

in geometric terms. There is at once therefore an immediate correspondence between the interaction of the vectors in real space (let us imagine them to be forces acting upon a particle) and the geometric portrayal of this interaction. If we now simplify the picture to the situation where we have only two forces acting upon a point and redraw the diagram such that $\mathbf{v}$ acts from A to B and $\mathbf{w}$ acts from B to C (figure 2.8), but retaining the lengths and directions of these two vectors, then taking each vector separately, $\mathbf{v}$ first moves the hypothetical particle from A to B, and $\mathbf{w}$ later moves it from B to C. The particle ultimately ends up at C, so that we now have a *resultant* vector $(\mathbf{v}+\mathbf{w})$ acting from A to C. This geometric construction is the *triangle law* of vectors. If we were now to double each vector quantity (but with no change in direction) it should be clear that we merely double the magnitude of the resultant vector. Thus:

$$\boldsymbol{v} + \boldsymbol{w} = (\boldsymbol{v} + \boldsymbol{w})$$

and

$$2\boldsymbol{v} + 2\boldsymbol{w} = 2(\boldsymbol{v} + \boldsymbol{w})$$

It is important at this early stage to note the difference between vector and ordinary arithmetic addition. As the reader should see the processes involved are entirely different in the geometric sense. However we can see that the multiplication of a vector by a scalar quantity produces another vector. Note however that if we were to multiply the vector by -1 we should reverse its direction without changing its magnitude, and produce the vector $-\boldsymbol{v}$. Thus a westerly wind can be depicted simply as a negative easterly wind. This has further important ramifications in the next section.

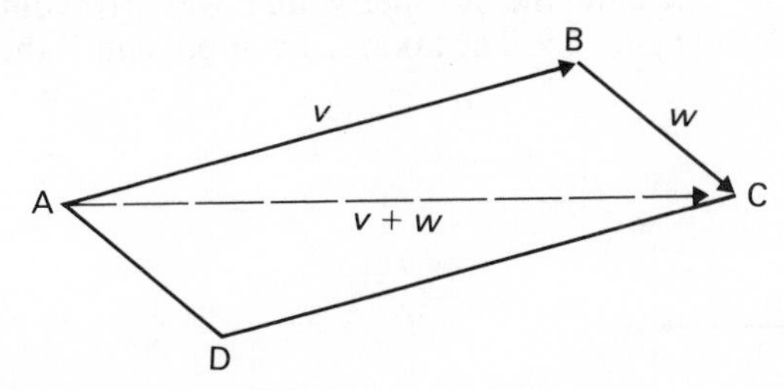

Figure 2.8 Triangle law and parallelogram.

One development of the triangle law is to produce a parallelogram ABCD by mirroring the triangle ABC in figure 2.8 about the resultant AC. We term the area contained within parallelogram ABCD, the *vector area* ($\boldsymbol{S}$) of the vectors $\boldsymbol{v}$ and $\boldsymbol{w}$, or:

$$\boldsymbol{S} = \boldsymbol{v} \wedge \boldsymbol{w} \tag{2.31}$$

From the information given in sections 2.1 and 2.2 on trigonometric ratios we can see also that:

$$S = \text{AB.BC} \sin(\text{angle ABC})$$

where S is the magnitude of $\boldsymbol{S}$, which again relates the vector magnitude directly to the geometry. The construction of a parallelogram around two vectors has considerable application in the treatment of forces acting at a point and the figure produced is called the *parallelogram of forces*.

2.5 Components

In the introduction to vectors in section 2.3 we have already referred to the taking of *components*, if not actually in name. The acceleration due to gravity acts vertically downwards, and when we are considering for example hillslope stability, one of the factors we have to bear in mind is the *vertical component* of g acting upon fragments of the slope surface. If the slope is nearly horizontal then clearly the total downward force acting upon pieces of rock or soil is very small indeed. Anything which has both magnitude and direction can be resolved in any direction. We can take horizontal or vertical, westerly or northerly or any other components resolved in whatever direction is necessary. If we are interested for example, in the overall 'run' of westerly winds for a site then a good start to analysis would be obtained by taking the westerly components of wind. So that if we had a wind of 5 m s^{-1} from due west this would have a westerly component of +5 m s^{-1}, and an easterly wind of similar magnitude would have a westerly component of −5 m s^{-1}. By the same token, winds from due north or south will have no westerly component at all.

Clearly what we are doing is the rationalization of a vector quantity into a standard form which is immediately comparable with other measures of the same phenomenon. This is often very convenient and the taking of components simply involves the use of the simple trigonometric ratios of sine and cosine. Consider a wind of velocity 10 m s^{-1} from the direction 30 degrees west of north, or in conventional terms 330 degrees clockwise from north. It is a simple matter to calculate the northerly and westerly components. The geometry is shown in figure 2.9. The taking of components is the reverse of the

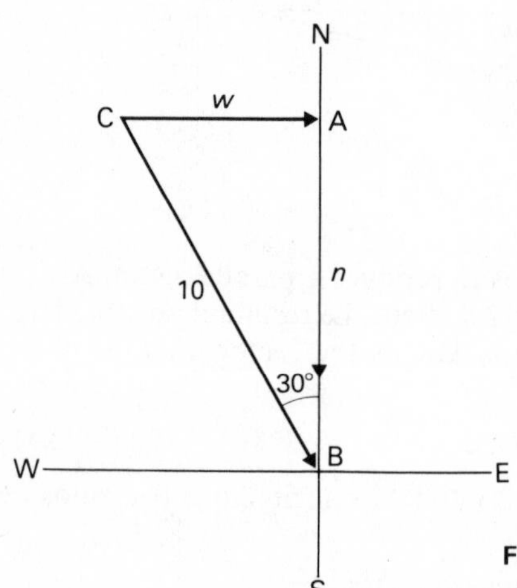

Figure 2.9 Components (*w* and *n*) of a wind of 10 m s^{-1} blowing from 330 degrees.

computation of the resultant, outlined in section 2.4. In this case however, the resultant (10 m s^{-1} from 330 degrees) is known, and we need to compute the two vectors acting at right-angles to one another which would produce this. Thus we can represent the northerly (n) and westerly (w) components as follows:

$$w = 10 \sin 30 = 5 \text{ m s}^{-1}$$

and

$$n = 10 \cos 30 = 8{\cdot}66 \text{ m s}^{-1}$$

Of course it similarly follows that $n^2 + w^2 = 10^2$, since trinagle ABC is right-angled.

This suggests that we can formalize vector components in more general terms. The situation we have considered so far is the most simple possible, utilizing only two dimensions. This has served to introduce the topic. Space is three-dimensional and therefore we can usefully extend our terms of reference to three coordinate axes, x, y and z, and treat vector relationships in general algebraic terms with respect to these planes. Let us take the vector $\boldsymbol{a}$ which can be defined in terms of its three mutually perpendicular components so that its end point is at point A (x,y,z). If we were defining the point A by means of the vector $\boldsymbol{a}$ then we should call $\boldsymbol{a}$ the *position vector*, since it would precisely indicate where A was in relation to the three axes. We can further define the vector $\boldsymbol{a}$ as:

$$\boldsymbol{a} = \boldsymbol{x}+\boldsymbol{y}+\boldsymbol{z}$$

At this stage it is convenient to introduce the *unit vector*. A unit vector is as its name suggests, the vector whose value is one, and for more advanced vector computation it is convenient to express vectors in terms of the unit vector. The symbol we shall use to indicate the unit vector of $\boldsymbol{a}$ is $\hat{\boldsymbol{a}}$. Thus the vector $\boldsymbol{a}$ can be defined as $a\hat{\boldsymbol{a}}$, or for clarity we can express the magnitude of the vector as $|\boldsymbol{a}|$. So if we were to call the unit vectors along the x, y and z axes respectively $\boldsymbol{i}, \boldsymbol{j}$ and $\boldsymbol{k}$, then the above equation can be rewritten:

$$\boldsymbol{a} = \boldsymbol{i}x+\boldsymbol{j}y+\boldsymbol{k}z \tag{2.32}$$

This relationship can be seen in figure 2.10 from which it should also be clear that:

$$|\boldsymbol{a}| = \sqrt{(x^2+y^2+z^2)} \tag{2.33}$$

by Pythagoras. Similarly the *direction* of $\boldsymbol{a}$ can be represented by the following equations:

$$x = |\boldsymbol{a}| \cos \alpha \tag{2.34}$$

$$y = |\boldsymbol{a}| \cos \beta \tag{2.35}$$

$$z = |\boldsymbol{a}| \cos \gamma \tag{2.36}$$

The cosines of α, β and γ are known as the *direction cosines* of the vector $\boldsymbol{a}$. Thus substituting in equation (2.32) for x, y and z we arrive at:

$$\hat{\boldsymbol{a}} = \frac{\boldsymbol{a}}{|\boldsymbol{a}|} = \boldsymbol{i} \cos \alpha+\boldsymbol{j} \cos \beta+\boldsymbol{k} \cos \gamma \tag{2.37}$$

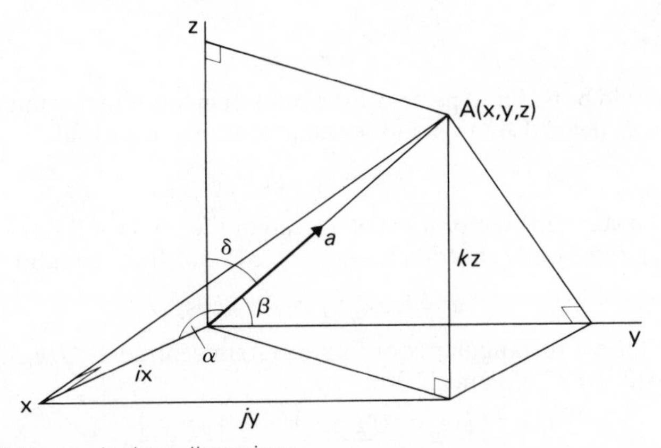

Figure 2.10 Vectors in three dimensions.

Problems 2.5

2.5.1 Resolve a southwesterly wind of 5·7 m s^{-1} into westerly and northerly components.

2.5.2 Magnitude of $\boldsymbol{v} = 3{\cdot}2$, magnitude of $\boldsymbol{w} = 2{\cdot}1$, enclosed angle = 45 degrees. Find the magnitude of the resultant $\boldsymbol{v}+\boldsymbol{w}$ and the angle it subtends with $\boldsymbol{w}$.

2.5.3 A rock rests on a hillslope whose slope is 1 in 7, what is the vertical component of the force acting on this rock if its weight is given by *mg*.

2.5.4 Find the resultant magnitude and direction of a wind produced by the combination of an onshore sea breeze of 14·0 m s^{-1} and a gradient wind of 9·0 m s^{-1} blowing offshore at 60 degrees to the coastline.

2.5.5 An updraught of 5 m s^{-1} exists within a part of a cumulus cloud which is subjected to a horizontal wind of 2·5 m s^{-1}. *Compute* the velocity of air movement in the cloud.

2.6 The straight line

We are now in a position to define an equation for a straight line in vector form. Suppose P is an arbitrary point on a straight line (figure 2.11), with Q being

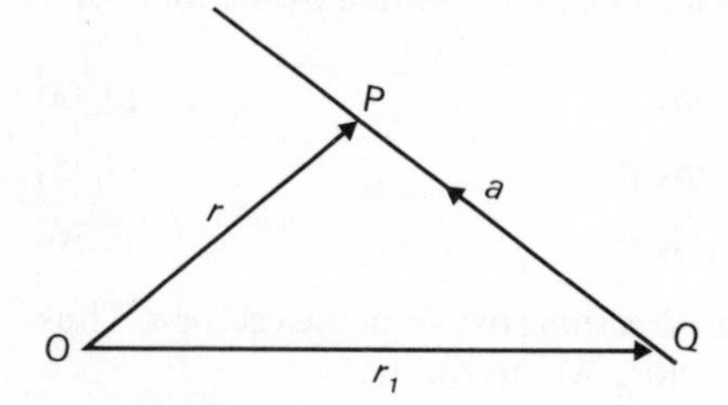

Figure 2.11 The straight line.

another point on the same line. If point O is the origin for the coordinate system and we let P have position vector $\boldsymbol{r}$, Q have position vector $\boldsymbol{r}_1$, and $\overrightarrow{QP} = \boldsymbol{a}$, then,

$$\boldsymbol{r} = \boldsymbol{r}_1 + \boldsymbol{a}$$

Again it would be better to portray the equation in terms of the unit vector $\hat{\boldsymbol{a}}$, so that QP was defined by this and a scalar multiple, *t*, so that:

$$\boldsymbol{r} = \boldsymbol{r}_1 + t\hat{\boldsymbol{a}} \tag{2.38}$$

which is the standard vector form of a straight line. Now if P has coordinates (x_1, y_1, z_1) and Q has coordinates (x_2, y_2, z_2) and from equation (2.37):

$$\boldsymbol{a} = \boldsymbol{i}\cos\alpha + \boldsymbol{j}\cos\beta + \boldsymbol{k}\cos\gamma$$

then if we have a rectangular coordinate system centred at O, we have, using equations (2.34), (2.35) and (2.36):

$$\frac{(x_2 - x_1)}{\cos\alpha} = \frac{(y_2 - y_1)}{\cos\beta} = \frac{(z_2 - z_1)}{\cos\gamma} = t \tag{2.39}$$

The expression of each of x, y and z in terms of t and the direction cosine is called the *parametric form* of the equation.

2.7 The scalar product

The next logical step is to see what results from the resolution of *two* vectors into one plane, or the *projection* of one upon the other. In figure 2.12 we see two

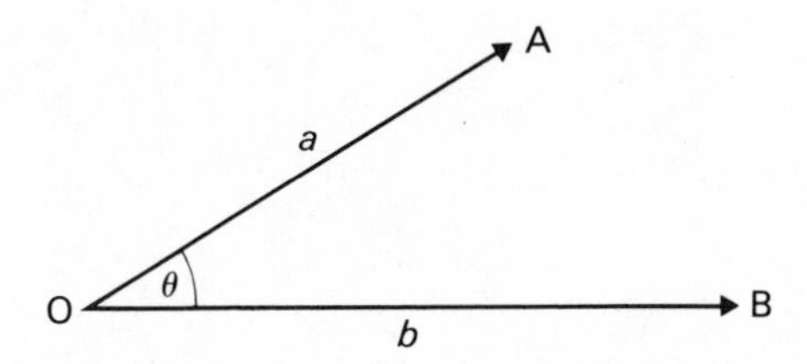

Figure 2.12 Scalar or dot product.

vectors $\boldsymbol{a}$ and $\boldsymbol{b}$, that is resolving $\boldsymbol{a}$ in the direction of $\boldsymbol{b}$ is given simply by the geometric expression:

$$|\boldsymbol{a}|\ |\boldsymbol{b}| \cos\theta \tag{2.40}$$

The projection of $\boldsymbol{b}$ upon $\boldsymbol{a}$ would similarly be $|\boldsymbol{b}|\ |\boldsymbol{a}| \cos\theta$, which is clearly the same magnitude as equation (2.40). This operation is known as the *scalar product* of two vectors, or as the *dot product*, since we write:

$$\boldsymbol{a}.\boldsymbol{b} = |\boldsymbol{a}||\boldsymbol{b}| \cos\theta \tag{2.41}$$

If $\boldsymbol{a}$ and $\boldsymbol{b}$ are each defined in terms of the three axes given previously then from equation (2.32):

$$\boldsymbol{a} = a_1\boldsymbol{i} + a_2\boldsymbol{j} + a_3\boldsymbol{k} \tag{2.42}$$

and

$$\boldsymbol{b} = b_1\boldsymbol{i} + b_2\boldsymbol{j} + b_3\boldsymbol{k} \tag{2.43}$$

Now if we multiply out these results to obtain $\boldsymbol{a}.\boldsymbol{b}$ we have:

$$\boldsymbol{a}.\boldsymbol{b} = a_1b_1 + a_2b_2 + a_3b_3 \tag{2.44}$$

since:

$$\boldsymbol{i}.\boldsymbol{i} = \boldsymbol{j}.\boldsymbol{j} = \boldsymbol{k}.\boldsymbol{k} = 1$$

as θ in these dot products is zero; and:

$$\boldsymbol{i}.\boldsymbol{j} = \boldsymbol{j}.\boldsymbol{k} = \boldsymbol{k}.\boldsymbol{i} = 0$$

as θ in these dot products is 90°.

2.8 The vector product

Whereas the scalar product of two vectors derives a value for the two combined in scalar terms by resolving one upon the other, the *vector product* of two vectors provides a combination which has both magnitude and direction—it is in other words, another vector. The vector product is defined as follows. If $\boldsymbol{a}$ and $\boldsymbol{b}$ are two vectors in the same plane, and θ is the angle between them measured from $\boldsymbol{a}$ to $\boldsymbol{b}$ (this as we shall see is very important) then the vector product is:

$$\boldsymbol{a} \wedge \boldsymbol{b} = |\boldsymbol{a}|\,|\boldsymbol{b}| \sin\theta\, \hat{\boldsymbol{c}} \tag{2.45}$$

where $\hat{\boldsymbol{c}}$ is the unit vector acting perpendicular to the plane of $\boldsymbol{a} \wedge \boldsymbol{b}$ (figure 2.13). The vector product of $\boldsymbol{a} \wedge \boldsymbol{b}$ is a vector of *magnitude* $|\boldsymbol{a}|\,|\boldsymbol{b}| \sin\theta$, and of direction

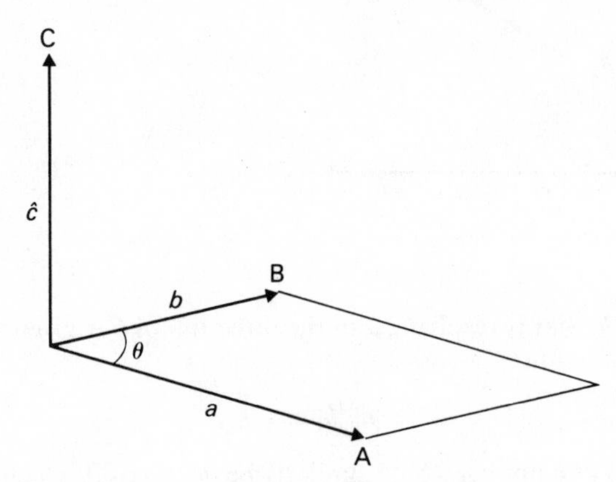

Figure 2.13 Vector product.

$\hat{\boldsymbol{c}}$. A parallelogram can be constructed about $\boldsymbol{a}$ and $\boldsymbol{b}$ so that the vector area (equation (2.31)) is also defined by $|\boldsymbol{a}|\,|\boldsymbol{b}| \sin\theta$. Note however, that the vector product of $\boldsymbol{b} \wedge \boldsymbol{a}$ is:

$$\begin{aligned} \boldsymbol{b} \wedge \boldsymbol{a} &= |\boldsymbol{b}|\,|\boldsymbol{a}| \sin(-\theta)\hat{\boldsymbol{c}} \\ &= |-|\boldsymbol{b}|\,|\boldsymbol{a}| \sin\theta\, \hat{\boldsymbol{c}} = -\boldsymbol{a} \wedge \boldsymbol{b} \end{aligned} \tag{2.46}$$

This should be obvious from the fact that we are now measuring the angle θ from $\boldsymbol{b}$ to $\boldsymbol{a}$. The result is therefore that whereas the scalar product was commutative, the vector product is not, and this should always be borne in mind in computation.

2.9 Computation with scalar and vector products

Suppose we wish to determine the angle θ between two vectors $\boldsymbol{a}$ and $\boldsymbol{b}$. This can be approached using both the scalar and vector product rules outlined in sections 2.7 and 2.8. Let:

$$\boldsymbol{a} = 2\boldsymbol{i} + \boldsymbol{j} + 3\boldsymbol{k}$$

and

$$\boldsymbol{b} = \boldsymbol{i} + 2\boldsymbol{j} - \boldsymbol{k}$$

in terms of the three axes x, y and z, where $\boldsymbol{i}, \boldsymbol{j}$ and $\boldsymbol{k}$ are again unit vectors along each axis in turn. For the scalar product since:

$$\hat{\boldsymbol{a}} = \boldsymbol{i} \cos\alpha + \boldsymbol{j} \cos\beta + \boldsymbol{k} \cos\gamma \text{ (from equation (2.37))}$$

and as $\boldsymbol{i}.\boldsymbol{j} = \boldsymbol{j}.\boldsymbol{k} = \boldsymbol{k}.\boldsymbol{i} = 0$, and $\boldsymbol{i}.\boldsymbol{i} = \boldsymbol{j}.\boldsymbol{j} = \boldsymbol{k}.\boldsymbol{k} = 1$, then:

$$\hat{\boldsymbol{a}}^2 = \cos^2\alpha + \cos^2\beta + \cos^2\gamma$$

Now, from equation (2.33):

$$|\boldsymbol{a}| = \sqrt{(2^2+1^2+3^2)} = \sqrt{14}$$

and

$$|\boldsymbol{b}| = \sqrt{(1^2+2^2+1^2)} = \sqrt{6},$$

and from equation (2.37):

$$\hat{\boldsymbol{a}} = \frac{\boldsymbol{a}}{|\boldsymbol{a}|} \text{ and } \hat{\boldsymbol{b}} = \frac{\boldsymbol{b}}{|\boldsymbol{b}|}$$

then:

$$\hat{\boldsymbol{a}} = \frac{(2\boldsymbol{i}+\boldsymbol{j}+3\boldsymbol{k})}{\sqrt{14}}$$

$$\hat{\boldsymbol{b}} = \frac{(\boldsymbol{i}+2\boldsymbol{j}-\boldsymbol{k})}{\sqrt{6}}$$

and from equation (2.41), $\hat{\boldsymbol{a}}.\hat{\boldsymbol{b}} = \cos\theta$, so that

$$\cos\theta = \frac{(2\boldsymbol{i}+\boldsymbol{j}+3\boldsymbol{k})(\boldsymbol{i}+2\boldsymbol{j}+\boldsymbol{k})}{\sqrt{14}\sqrt{6}}$$

$$= \frac{(2+2-3)}{\sqrt{84}}$$

$$= \frac{1}{\sqrt{84}}$$

$$= 0{\cdot}1091$$

and so $\theta = 83°\,44'$ (from tables)

Now, using the vector product:

$$\hat{\boldsymbol{a}} \wedge \hat{\boldsymbol{b}} = \frac{(2\boldsymbol{i}+\boldsymbol{j}+3\boldsymbol{k})}{\sqrt{14}} \wedge \frac{(\boldsymbol{i}+2\boldsymbol{j}-\boldsymbol{k})}{\sqrt{6}} \tag{2.47}$$

But for the vector product:

$$\boldsymbol{i}\wedge\boldsymbol{i} = \boldsymbol{j}\wedge\boldsymbol{j} = \boldsymbol{k}\wedge\boldsymbol{k} = 0,$$

since the included angle has the sine zero. However, since the three axes are mutually at right-angles where $\sin\theta = 90$ degrees, then:

$$\boldsymbol{i}\wedge\boldsymbol{j} = \boldsymbol{k} \text{ and } -\boldsymbol{j}\wedge\boldsymbol{i} = \boldsymbol{k} \tag{2.48}$$

$$\boldsymbol{j}\wedge\boldsymbol{k} = \boldsymbol{i} \text{ and } -\boldsymbol{k}\wedge\boldsymbol{j} = \boldsymbol{i} \tag{2.49}$$

$$\boldsymbol{k}\wedge\boldsymbol{i} = \boldsymbol{j} \text{ and } -\boldsymbol{i}\wedge\boldsymbol{k} = \boldsymbol{j} \tag{2.50}$$

The reader is urged very strongly to memorize these vector products and note that a change in sign of *one* vector in each couple will induce a change in sign of the resultant, and that a change in the sign of *both* results in the same resultant.

A reversal in the order of sequence (say, from $\boldsymbol{i} \wedge \boldsymbol{j}$ to $\boldsymbol{j} \wedge \boldsymbol{i}$) also changes the sign of the resultant. So in our example the vector product $\hat{\boldsymbol{a}} \wedge \hat{\boldsymbol{b}}$ becomes:

$$\hat{\boldsymbol{a}} \wedge \hat{\boldsymbol{b}} = \frac{(2\boldsymbol{i}+\boldsymbol{j}+3\boldsymbol{k}) \wedge (\boldsymbol{i}+2\boldsymbol{j}-\boldsymbol{k})}{\sqrt{84}}$$

$$= \frac{4(\boldsymbol{i} \wedge \boldsymbol{j})+2(\boldsymbol{i} \wedge -\boldsymbol{k})+(\boldsymbol{j} \wedge \boldsymbol{i})+(\boldsymbol{j} \wedge -\boldsymbol{k})+3(\boldsymbol{k} \wedge \boldsymbol{i})+6(\boldsymbol{k} \wedge \boldsymbol{j})}{\sqrt{84}}$$

$$= \frac{4\boldsymbol{k}+2\boldsymbol{j}+3\boldsymbol{j}-\boldsymbol{k}-\boldsymbol{i}-6\boldsymbol{i}}{\sqrt{84}}$$

$$= \frac{-7\boldsymbol{i}+5\boldsymbol{j}+3\boldsymbol{k}}{\sqrt{84}}$$

now since by definition $\hat{\boldsymbol{a}} \wedge \hat{\boldsymbol{b}} = \sin\theta\,\hat{\boldsymbol{c}}$, then

$$|\hat{\boldsymbol{a}} \wedge \hat{\boldsymbol{b}}| = |\sin\theta|$$

and therefore:

$$\sin^2\theta = \frac{(-7\boldsymbol{i}+5\boldsymbol{j}+3\boldsymbol{k}).(-7\boldsymbol{i}+5\boldsymbol{j}+3\boldsymbol{k})}{84}$$

(note here that we have a dot product) so

$$\sin^2\theta = \frac{49+25+9}{84}$$

$$\sin\theta = 0{\cdot}9940$$
$$= 83^\circ\,44'$$

Problems 2.9

Find the dot ($\boldsymbol{a}.\boldsymbol{b}$) and vector ($\boldsymbol{a} \wedge \boldsymbol{b}$) products of the following vectors defined in terms of $\boldsymbol{i}$, $\boldsymbol{j}$ and $\boldsymbol{k}$, together with the value of θ.

2.9.1 $\boldsymbol{a} = 5\boldsymbol{i}+2\boldsymbol{j}+\boldsymbol{k}$
$\boldsymbol{b} = 3\boldsymbol{i}+5\boldsymbol{j}+2\boldsymbol{k}$

2.9.2 $\boldsymbol{a} = 2\boldsymbol{j}-6\boldsymbol{k}$
$\boldsymbol{b} = 2\boldsymbol{i}-3\boldsymbol{j}-\boldsymbol{k}$

2.9.3 $\boldsymbol{a} = 5{\cdot}2\boldsymbol{i}-3{\cdot}6\boldsymbol{j}+0{\cdot}1\boldsymbol{k}$
$\boldsymbol{b} = -9{\cdot}7\boldsymbol{i}+3{\cdot}1\boldsymbol{j}+10{\cdot}2\boldsymbol{k}$

3
Linear functions

The concept of a mathematical function was introduced in chapter 1. A function defines the proportionality between two variables, for example in the general sense, between x and y. It does this in two ways, firstly by signifying whether the proportionality is a constant ratio between x and y for all values of each, and secondly by defining the magnitude of this ratio. For example in the function $y = 3x$, the ratio of y to x is always 3:1, but in the function $y = 3x^2$ the ratio between the two variables increases as x increases. In this chapter we are concerned solely with the first type of function, the *linear function.*

3.1 The basic linear function

The general equation for a linear function relating y and x ($y = mx + k$) was introduced in chapter 1, and its expression in vector form was deduced in the previous chapter. For the present we shall assume that $k = 0$, and thus that x and y are always related in the constant ratio $1{:}m$, indicating that for unit change in x, y changes value by m. Taking successive values for x of 1, 2 and 3, y will equal $1m$, $2m$ and $3m$. If $m = 2$ then coordinate pairs of x and y will be in the ratio 1:2, for example, (1,2), (2,4) and (3,6). If these points are plotted with respect to the axes of a graph, then the line joining them will be straight and will pass through the origin (0,0). This is an elementary linear function $y = 2x$ (figure 3.1), and together with all other linear functions where $k = 0$, it reflects a constant ratio between x and y. Clearly however, if the function had been $y = 2x^2$, then, whilst for example it would share point (0,0) with $y = 2x$, the ratio between x and y when $x = 1$ would be 1:2, when $x = 2$, 1:4, when $x = 3$, 1:6 and so on. This non-linear function is also shown in figure 3.1.

Many relationships in physical geography can be portrayed as linear functions if only as approximations. For example the equation $q = 5{\cdot}1a^{0{\cdot}98}$ cited in chapter 1, was very nearly linear. As we saw in that chapter the linear function of $q = 5{\cdot}1a$ very closely approximated it, merely by increasing terms in a by a factor of $a^{0{\cdot}02}$ ($a^{1{\cdot}0} = a^{0{\cdot}98}a^{0{\cdot}02}$). The term m in the general equation besides indicating the magnitude of the ratio between q and a or y and x, is also the tangent of the angle between the linear function when plotted on a graph, and the x-axis. Hence it is referred to as the *slope* or the *gradient* of the line. In equations where this term is very small therefore the graph of the function is more nearly parallel to the x-axis (e.g. $y = 0{\cdot}2x$ in figure 3.1). Similarly, very large values of m indicate that the graph of the function is more nearly at right-angles to the x-axis (e.g. $y = 20x$ in figure 3.1). Since the tangent of 90 degrees is infinity, the equation of a line parallel to the y-axis is given by $y = \infty x$, or $x = \frac{y}{\infty} = 0$. Reference to figure 2.5 will further indicate that, whilst the graphs of

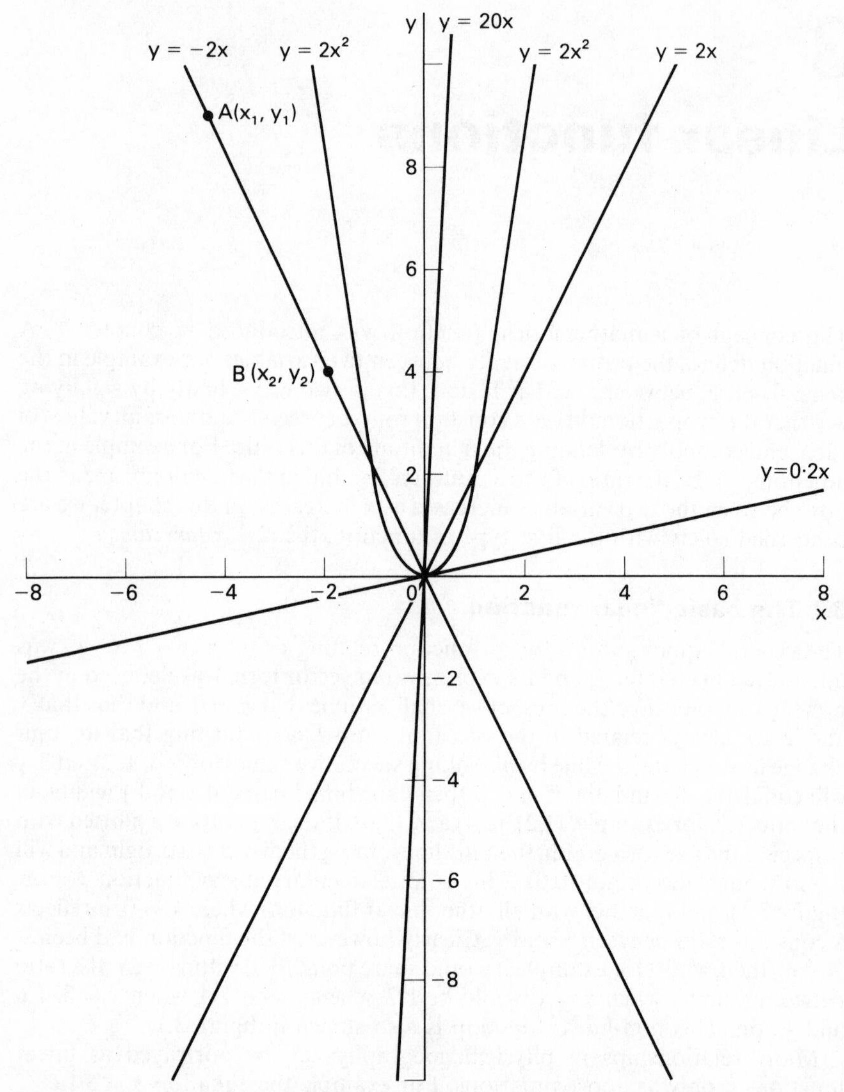

Figure 3.1 Simple linear functions: $y = -2x$, $y = 0{\cdot}2x$, $y = 2x$, $y = 20x$, and the non-linear function $y = 2x^2$.

functions which slope upwards from the left will have positive gradients, those which slope in the opposite direction must have negative gradients, for example, $y = -2x$ in figure 3.1. The gradient is thus the *rate of change* of one variable with another, or between any two points on the graph of a linear function of coordinates (x_1,y_1) and (x_2,y_2), the slope is:

$$m = \frac{y_2 - y_1}{x_2 - x_1} \tag{3.1}$$

The solution to problem 1.8.1 will have suggested this to many readers. Thus by taking any two points A (x_1,y_1) and B (x_2,y_2) on the graph of $y = -2x$ in figure 3.1 where A = (−4·5,9·0) and B = (−2·0,4·0), we can see that:

$$m = \frac{9{\cdot}0 - 4{\cdot}0}{(-4{\cdot}5) - (-2{\cdot}0)} = \frac{5{\cdot}0}{-2{\cdot}5} = -2$$

The graphs of all linear functions in which $k = 0$ in the general form $y = mx + k$, pass through the origin of the axes at (0,0). By varying the value of k where $k \neq 0$ we can obtain a 'family' of parallel straight lines, all of which have slope m. In figure 3.2 graphs have been drawn for $y = 2x + k$ for $k = -2$ to $+3$. From these it is clear that the value of k determines directly the point at which the graph of the function crosses the y-axis. The constant k is thus termed the *intercept*, and for the general equation the intersection between the line and the y-axis occurs at a point $(0,k)$. The relationship linking Fahrenheit and Celsius temperatures used in problem 1.8.2 is a convenient and well known example of a linear function with which to work. The ratio between one degree Fahrenheit and one degree Celsius is 9:5, since there are 180 graduations between freezing point and boiling point on the Fahrenheit scale (212–32 degrees), and 100

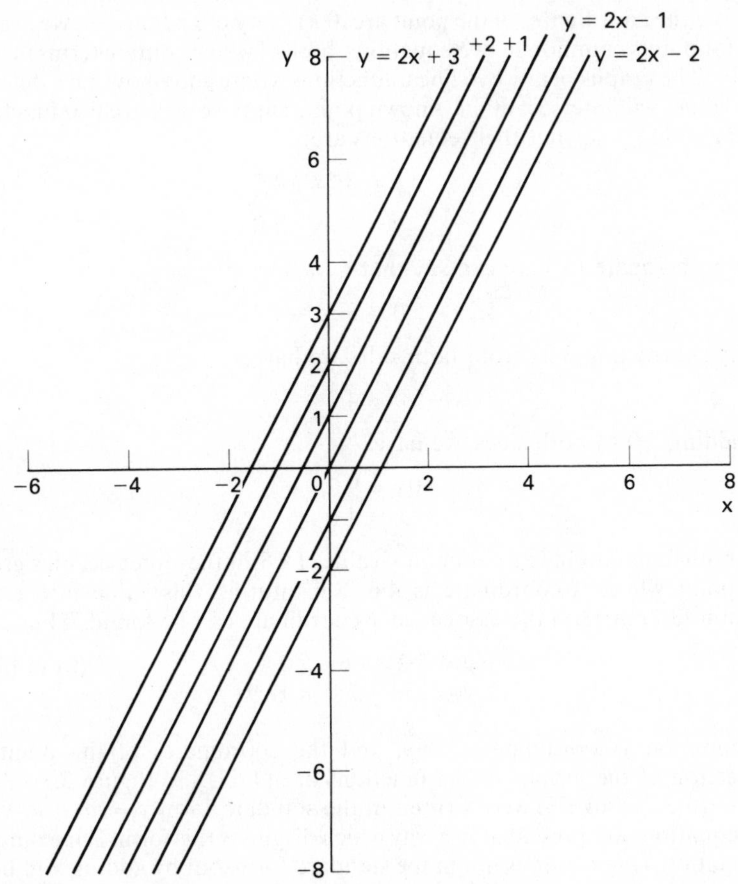

Figure 3.2 Family of linear functions: $y = 2x + k$, $k = -2$ to $+3$.

between the same points on the Celsius scale. Therefore the function relating F to C has slope, m, equal to 1·8, so that:

$$F = 1{\cdot}8C + k$$

However, we also know that freezing point is represented by $C = 0$ and by $F = 32$, and therefore substituting in $F = 32$ and $C = 0$ in our equation we obtain $k = 32$ and the final linear equation is $F = 1{\cdot}8C + 32{\cdot}0$. The reader is now strongly advised to familiarize himself with the plotting of linear functions by inventing suitable values of m and k. He is also advised to try the reverse process: that of drawing a number of straight lines on graph paper and obtaining their equations.

3.2 Simultaneous equations

The intercept conveniently provides us with the coordinates of a point where the line intersects with the y-axis. In effect to determine this point the two equations $y = mx + k$ and $x = 0$ are being solved *simultaneously* to find the coordinates of the point common to both. In this case the solution is produced simply by substituting $x = 0$ into the general equation, so that $y = k$, and therefore the coordinates of the point are $(0, k)$. Very often however, we need to solve for points common to two equations, both of which contain terms in both x and y. The graphs of any two linear functions whose equations have different slope values will intersect at *one* known point. Suppose we have two functions relating x and y so that their equations are:

$$y = 3{\cdot}2x + 4{\cdot}2 \qquad (3.2)$$

and

$$y = 5{\cdot}2x - 5{\cdot}0 \qquad (3.3)$$

Since both equate to y we can say that:

$$5{\cdot}2x - 5{\cdot}0 = 3{\cdot}2x + 4{\cdot}2$$

and by subtracting $3{\cdot}2x$ from both sides we have,

$$2{\cdot}0x - 5{\cdot}0 = 4{\cdot}2$$

and adding 5·0 to both sides, we have,

$$2{\cdot}0x = 9{\cdot}2,$$

and

$$\therefore\ x = 4{\cdot}6$$

Thus both functions have a common x-value of 4·6, or they intersect on a graph at a point whose x-coordinate is 4·6. Substituting this value into either equation (3.2) or (3.3) the associated y-coordinate can be found. Thus,

$$y = 5{\cdot}2 \times 4{\cdot}6 - 5{\cdot}0 \qquad \text{(from (3.3))}$$

and

$$\therefore\ y = 23{\cdot}9 - 5{\cdot}0 = 18{\cdot}9$$

The common y-coordinate is 18·9, and the coordinates of the point of intersection of the graphs of the functions are (4·6, 18·9) (figure 3.3). Both equations (3.2) and 3.3) were written in the standard form $y = mx + k$. Very often equations are presented in a way which disguises this form. For example, the equation $4y - x = 16$ is not in the standard form, but by adding x to both sides of the equation we obtain $4y = x + 16$, and dividing through by four:

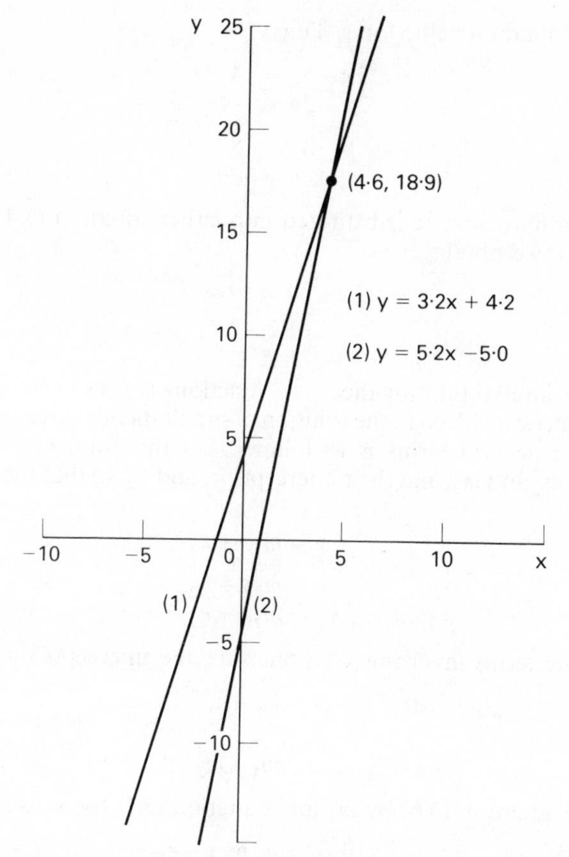

Figure 3.3 Graphical solution of $y = 3{\cdot}2x + 4{\cdot}2$ and $y = 5{\cdot}2x - 5{\cdot}0$.

$y = \frac{1}{4}x + 4$, so that $m = \frac{1}{4}$ and $k = 4$. It is often desirable, though with practice not essential, when attempting to solve simultaneous equations, to manipulate the terms in each equation such that all terms in x and y are grouped in the same order in both equations. Consider a second example below, where equations (3.4) and (3.5) must be solved simultaneously.

$$2x - 5y = 14 \tag{3.4}$$

$$x = y - 2 \tag{3.5}$$

Regrouping (3.5) to conform with (3.4) we have:

$$2x - 5y = 14$$
$$x - y = -2$$

We must now eliminate either x or y by combining both equations. Regrouping the second equation again we could obtain the equation in standard form and substitute in (3.4) for y, but this is rather laborious. Instead if we multiply the second equation by two and then subtract it from the first equation we obtain a value for y. Thus:

$$\begin{array}{lr} & 2x-5y = 14 \\ & 2x-2y = -4 \\ \hline \text{subtracting} & 0-3y = 18 \\ & \therefore \underline{y = -6} \end{array}$$

This solution may now be substituted into either equation (3.4) or (3.5). In equation (3.4) we obtain:

$$\begin{aligned} 2x+30 &= 14 \\ \therefore 2x &= -16 \\ \therefore \underline{x} &\underline{= -8} \end{aligned}$$

The point of intersection for these two functions is thus $(-8, -6)$.

The arithmetic involved in the solution of simultaneous equations expressed in even more general terms is as follows. Let the gradients of two linear functions be m_1 and m_2, and their intercepts k_1 and k_2, so that the equations of the two are:

$$y = m_1x+k_1 \tag{3.6}$$

$$y = m_2x+k_2 \tag{3.7}$$

then:

$$m_1x+k_1 = m_2x+k_2$$

and gathering terms involving x on one side and intercepts on the other,

$$x(m_1-m_2) = k_2-k_1$$

and thus:

$$x = \frac{k_2-k_1}{m_1-m_2} \tag{3.8}$$

Multiplying equation (3.6) by m_2 and equation (3.7) by m_1 we have:

$$m_2y = m_2m_1x+m_2k_1 \tag{3.9}$$

and

$$m_1y = m_1m_2x+m_1k_2 \tag{3.10}$$

Subtracting (3.10) from (3.9) and gathering terms on the left-hand side we have:

$$y(m_2-m_1) = m_2k_1-m_1k_2$$

$$\therefore y = \frac{m_2k_1-m_1k_2}{m_2-m_1} \tag{3.11}$$

Hence the coordinates of the point of intersection of the graphs of any two linear functions are:

$$\left(\frac{k_2-k_1}{m_1-m_2}, \frac{m_2k_1-m_1k_2}{m_2-m_1}\right) \tag{3.12}$$

We are now in a position to apply the mathematics of linear functions to an elementary problem in physical geography. Many may have difficulty in understanding and interpreting relationships between the various atmospheric lapse rates, and it is with these problems in mind that the following example has been included. The degree to which the atmosphere is stable or unstable is determined by the moisture content of the air and its degree of saturation, the temperature lapse rate upwards through the atmosphere (the environmental lapse rate—ELR) and the surface temperature. Further amplification of the theory behind lapse rates and the gas laws in general is found in Davidson (1978). An unsaturated air parcel which is heated to a temperature greater than its surroundings will rise through the atmosphere at a rate determined by its buoyancy (itself a function of the temperature difference between the air parcel and its surroundings), and will cool at the dry adiabatic lapse rate (DALR) of 0·98°C/100 m of altitude as long as the air in the parcel remains unsaturated. A function relating temperature to altitude under unsaturated conditions will thus have a slope (m) of $-0{\cdot}98$. The temperature (T) at a given height (z m $\times 10^2$) for an unsaturated parcel is determined by this slope and the initial starting temperature (T_0) at the ground where $z = 0$, which is thus equivalent to the intercept. The complete function relating T to z is thus:

$$T = -0{\cdot}98z + T_0 \tag{3.13}$$

If the ground temperature is 20·0 °C, then at height z = 800 m, the temperature of the air parcel is:

$$T = -0{\cdot}98 \times 8 + 20{\cdot}0 = 12{\cdot}2\,°\mathrm{C}$$

The equation is valid as long as the air parcel remains unsaturated. Once it becomes saturated the release of latent heat as the parcel continues to rise partly compensates for the temperature fall with altitude, and thus a second function involving the saturated adiabatic lapse rate (SALR) must be introduced. The SALR is not a constant lapse—that is, the function which represents it is non-linear—as the amount of latent heat liberated is dependent upon the absolute moisture content of the parcel and therefore also upon its temperature. For convenience however, we shall assume the SALR to be constant at −0·50 °C/100 m, its value at about 1,000 m altitude and 10 °C. Again there is the need to establish a starting temperature in order to define fully the SALR function:

$$T = -0{\cdot}50z + k \tag{3.14}$$

The altitude at which the air parcel adopts the SALR instead of the DALR is assumed to be the height at which the original dew point temperature is reached. Thus if the surface temperature was 20·0 °C and the dew point 11·0 °C, this altitude can be determined by substituting the dew point value into the equation (3.13) for $T_0 = 20{\cdot}0$:

$$T = 20{\cdot}0 - 0{\cdot}98z \tag{3.15}$$

$$11{\cdot}0 = 20{\cdot}0 - 0{\cdot}98z \tag{3.16}$$

and

$$0{\cdot}98z = 9{\cdot}0$$

$$z = 9{\cdot}2 \text{ (or 920 } m\text{)}$$

This height marks the point at which the rising air parcel cooling at the DALR will become saturated, the adiabatic condensation level, and is the point from which the SALR equation (3.14) is valid. The SALR function will thus have $T = 11{\cdot}0$ when $z = 9{\cdot}2$ and therefore substituting in equation (3.14) we have:

$$11{\cdot}0 = -4{\cdot}6 + k$$

$$k = 15{\cdot}6$$

and thus the complete SALR equation is:

$$T = -0{\cdot}50z + 15{\cdot}6 \qquad (3.17)$$

The relationships between equations (3.14), (3.16) and (3.17) are shown graphically in figure 3.4. Axes are reversed as is convention in meteorology, so that height (z) the independent variable appears on the ordinate. This merely permits easier graphical interpretation.

The environmental lapse rate has yet to be introduced in mathematical terms. So far an assumption has been made that the atmosphere has been unstable: that is, the temperature of the air within the rising parcel is always greater than that in the surrounding air. In the real atmosphere the rate of change of temperature with height is never constant through all levels, and any function relating T to z would be extremely complicated and rarely linear. However, let us assume as we did for the SALR, that the trend can be approximated to a satisfactory degree by the expression:

$$T = -0{\cdot}30z + 13{\cdot}0 \qquad (3.18)$$

indicating a surface temperature of 13·0 °C and temperature lapse of 0·3 °C/100 m. This is also shown in figure 3.4. The height at which an air parcel heated to 20·0 °C at the surface would cease to rise will therefore be given by a simultaneous solution of equation (3.18) with either (3.14) (unsaturated) or (3.17) (saturated), since their intersections mark the point at which

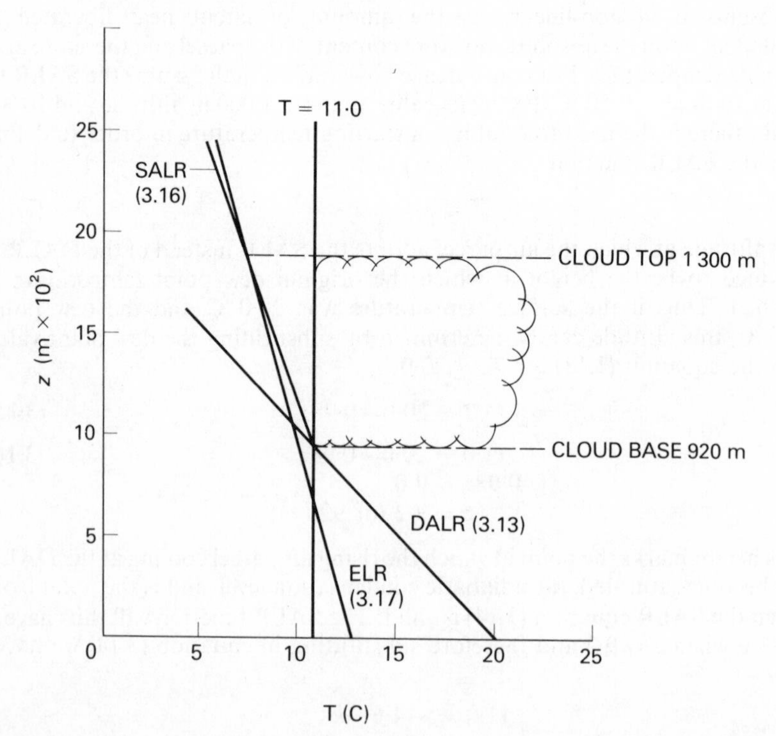

Figure 3.4 Simultaneous solution of lapse rates.

temperatures are the same in both the rising air parcel and the surrounding air. Solving equations (3.14) and (3.18) simultaneously we have:

$$20{\cdot}0 - 0{\cdot}98z = 13{\cdot}0 - 0{\cdot}30z$$
$$\therefore\ 0{\cdot}68z = 7{\cdot}0$$
$$\therefore\ z = 10{\cdot}29 \text{ or } 1{,}029\text{ m}$$

Since however, this is above the height at which the parcel would have become saturated (920 m) we must instead solve (3.17) and (3.18) simultaneously as follows:

$$15{\cdot}60 - 0{\cdot}50z = 13{\cdot}00 - 0{\cdot}30z$$
$$\therefore\ 0{\cdot}20z = 2{\cdot}60$$
$$\therefore\ z = 13{\cdot}00 \text{ or } 1{,}300\text{ m}$$

Thus under the conditions stated the air parcel will cease to rise at a height of 1,300 metres, which will also mark the height of the top of cumiliform cloud development, giving a cloud depth of 380 metres (1,300–920 m).

Problems 3.2

Solve the following simultaneous equations and check the results graphically:

3.2.1 $y = 3x - 4$
$y = 12x + 1$

3.2.2 $y = 1 - x$
$17x - 3y = 2$

3.2.3 $5x - 2y = 0$
$x = 2y - 2$

3.2.4 $x + y = 0$
$x - y = 1$

3.2.5 $2y = 5x + 2$
$y = 6$

Plot the following functions on a graph to determine points of intersection and check the results by the simultaneous solution of their equations:

3.2.6 $y = 12x + 2$
$y = 2 - 12x$

3.2.7 $\frac{1}{2}y = x + 1$
$2y = 4x + 5$

3.2.8 $2x + y = 0$
$4x + y = 2$

3.3 Rotation of axes

The theory established thus far is sufficient for most mathematical and statistical purposes. However, certain statistical methods (for example, principal components analysis, chapter 4) entail the rotation of axes to 'transform' equations. Consider a point P (figure 3.5) which has coordinates (x,y) with reference to axes X and Y, and coordinates (x_1,y_1) with reference to axes X_1 and Y_1. Respective X and Y axes are inclined at an angle θ to one another. Both axes have a common origin, so that OP is of length r for both.

OP subtends an angle ϕ with axis X_1 and $(\theta+\phi)$ with X. Now,

$$x = r\cos(\theta+\phi)$$

$\therefore$ $$x = r\cos\theta\cos\phi - r\sin\theta\sin\phi \text{ (from (2.13))}$$

but $$x_1 = r\cos\phi \quad (3.19)$$

and $$y_1 = r\sin\phi \quad (3.20)$$

and $\therefore$ $$x = x_1\cos\theta - y_1\sin\theta \quad (3.21)$$

and $$y = r\sin(\theta+\phi)$$

$\therefore$ $$y = r\sin\theta\cos\phi + r\cos\theta\sin\phi \text{ (from (2.12))}$$

so that, $$y = x_1\sin\theta + y_1\cos\theta \text{ (from (3.19) and (3.21))} \quad (3.22)$$

If OP in figure 3.5 has the equation $y = 5x$ then if the axes are rotated through 30 degrees, so that $\sin\theta = 0{\cdot}5000$ and $\cos\theta = 0{\cdot}8660$ (from tables in appendix 4), the equation relative to the new axes is:

$$x\sin\theta + y\cos\theta = 5(x\cos\theta - y\sin\theta) \text{ (from (3.21) and (3.22))}$$
$$0{\cdot}5x + 0{\cdot}8660y = 5\times0{\cdot}8660x - 5\times0{\cdot}5y$$
$$0{\cdot}8660y + 2{\cdot}5y = 4{\cdot}33x - 0{\cdot}5x$$
$$\therefore\ 3{\cdot}366y = 3{\cdot}83x$$
$$\therefore\ y = 1{\cdot}14x$$

Had the original equation not passed through the origin, then the new intercept (k_1) would be given by:

$$k_1 = k\cos\theta$$

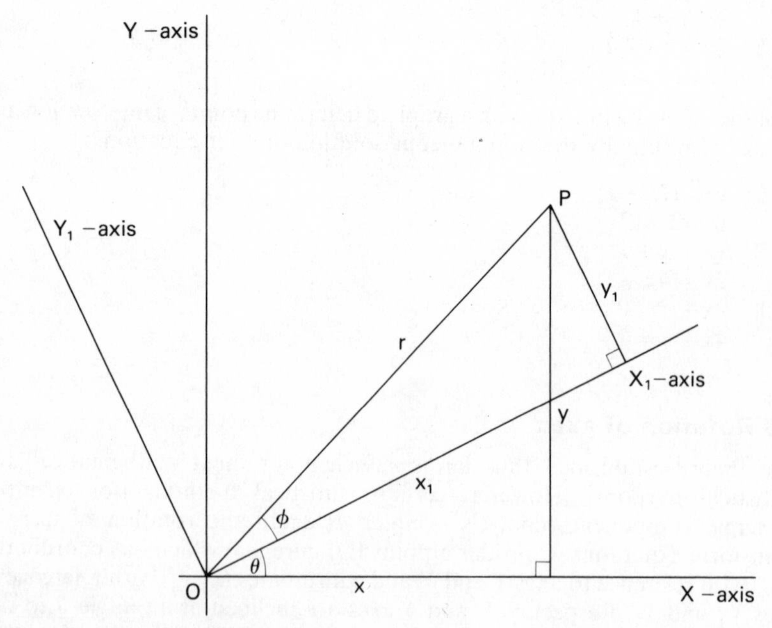

Figure 3.5 Rotation of axes.

So that if the equation had been $y = 5x+2$, transformation through angle 30 degrees would have yielded:

$$\begin{aligned} y &= 1{\cdot}14x+2\cos\theta \\ &= 1{\cdot}14x+1{\cdot}73 \end{aligned}$$

Transformations can be applied in the same way for non-linear functions, by substituting in equations (3.21) and (3.22).

3.4 Mathematical functions and statistical relationships

With the mathematical principles behind linear functions established we can now turn our attention towards the more general application of linear functions within physical geography. A function such as that relating temperature to altitude in defining the DALR represents a unique link established by physical laws, relating the two parameters T and z. This relationship can be arrived at theoretically and can be stated with absolute precision (see for example, McIntosh and Thom 1969, Davidson 1978). Since it is such a relationship we could use it to establish z for a given T as well as in the normal way of predicting T from z, although logically of course it is a simple conditional statement where z implies T. This relationship is a true mathematical function which is exact and unambiguous.

As mentioned in chapter 1 though, within the environmental sciences relationships frequently cannot be established to this degree of precision. Very often we are using inductive rather than deductive reasoning and rely upon relatively imprecise and inaccurate field measurement. In attempting to construct a mathematical model in the form of a function the best that can be attained is an approximation to the real relationship. In the relationship $q = 5{\cdot}1a^{0{\cdot}98}$ relating discharge to drainage area cited in chapter 1, variation in a did not explain all variation in q. Other factors also influence discharge and in addition, the equation was derived using field observation, where errors are almost inevitable. The means by which a mathematical function is arrived at which best describes the field information depends upon the sophistication of the data, the use to which the model is to be put and also upon the nature of the relationship (that is whether or not any degree of causality can be established by logical argument). If, as in most geographical relationships, causality can be established but the relationship is not a perfect one, then the definition of the dependent and independent variables becomes *statistically* important. It will be statistically invalid to use a relationship derived to predict y from x in order to predict x from y. The reasons for this will become apparent later in the chapter.

The need to be able to apply a 'best fit' function to an observed relationship raises two problems, and these apply equally to non-linear and multivariate situations. Firstly, an equation of best fit must be arrived at to provide the closest possible explanation for all observations. Secondly, as a result of this lack of correspondence between the function and the observed points, a measure of our *confidence* in the relationship needs to be established if the relationship is to be used for predictive purposes. Several statistical techniques exist to satisfy one or both of these requirements, and from this point on in the book reference is made to statistical techniques. It is outside the scope of this book to provide a detailed description and derivation of these methods and statistical formulae are given for guidance in appendix 5. The reader is however referred to a good elementary statistical text, such as Gregory (1970), Bryant (1966) or Siegel (1956) if he wishes to pursue the statistical side in greater depth.

3.5 Line fitting

One very simple way of determining a mathematical function which approximately describes a relationship between two variables is simply to plot the data pairs on a graph and to draw in a line by eye which appears best to represent all the data. Unfortunately there is no reliable way of assessing the goodness of fit of the line to the data points using this method, and it is therefore highly subjective. However, with experience, the method can produce lines from which the derived functions very closely approximate those determined by other, statistical, methods. This is particularly so when the scatter of points about the line is small—that is the statistical correlation (see appendix 5.2) is good. In drawing the line the aim should be to minimize the total deviation of all points from the line. An example of a line drawn in by eye is shown in figure 3.6. The plotted data points on the graph represent the general trend of decreasing mean annual temperature (Celsius) with altitude ($m \times 10^2$) for 35 climatological stations in Tanzania (East African Meteorological Department, 1964). The overall correlation appears good. The equation for the line can be determined by taking any two points on the line,

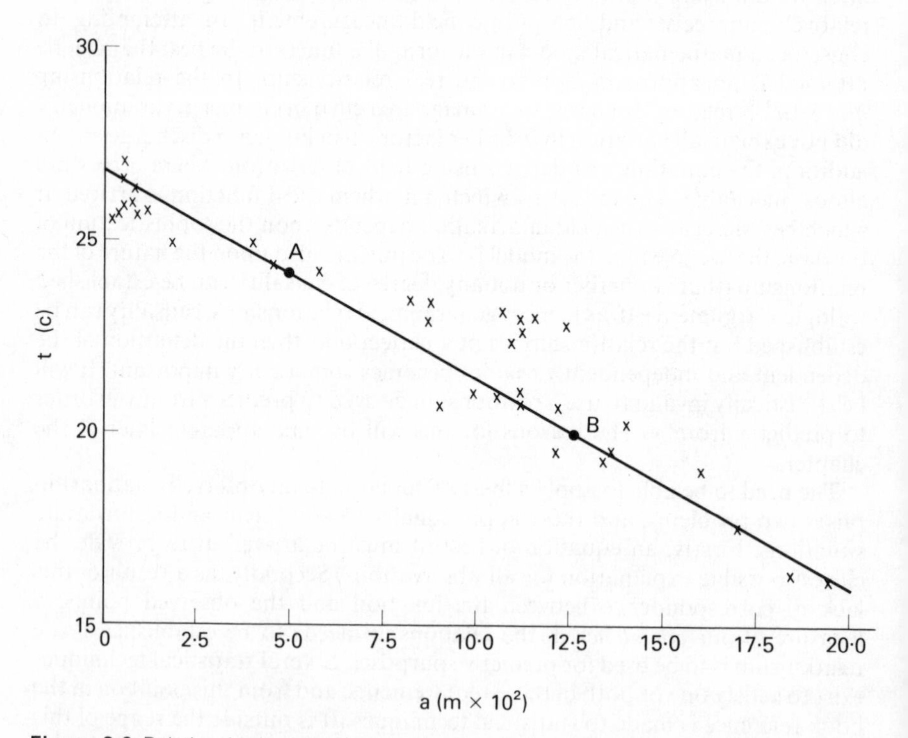

Figure 3.6 Relationship between mean annual temperature t (°C) and altitude a (10^2m) for Tanzania. Line fitted by eye.

for example A (5·0, 24·2) and B (13·0, 20·0), and substituting these into the general equation $y = mx + k$:

$$24{\cdot}2 = 5{\cdot}0m + k \tag{3.23}$$

$$20{\cdot}0 = 13{\cdot}0m + k \tag{3.24}$$

Subtracting (3.24) from (3.23) to remove k we have:

$$4{\cdot}2 = -8{\cdot}0m$$

$$m = -0{\cdot}53 \text{ (or } -0{\cdot}5 \text{ to one decimal place)}$$

It would be unrealistic to infer too great an accuracy into the method by using more than one decimal place in this example. Substituting this value into equation (3.23) we have:

$$24{\cdot}2 = 5{\cdot}0 \times -0{\cdot}5 + k$$

$$k = 24{\cdot}2 + 2{\cdot}5 = 26{\cdot}7$$

The derived equation is therefore:

$$\underline{t = 26{\cdot}7 - 0{\cdot}5a} \tag{3.25}$$

Alternatively we could have noted that the value for t when $a = 0$ (the intercept) was 26·9 and since m is given by $(y_1 - y_0)/(x_1 - x_0)$ (from equation (3.1)), then the function relating t to a is:

$$t = \left(\frac{20{\cdot}0 - 24{\cdot}2}{13{\cdot}0 - 5{\cdot}0}\right)a + 26{\cdot}9$$

$$\underline{t = 26{\cdot}9 - 0{\cdot}5a} \tag{3.26}$$

Note that equations (3.25) and (3.26) differ. This reflects the general inaccuracy of deriving functions using this technique. The method has the benefit of comparative speed, but there is no statistical objectivity. We have no way of being certain that the derived equation is in fact the one which best fits the data points, nor can we express any degree of confidence in predicted values of t from a.

A more precise derivation of a function can be obtained using statistical methods. The application of statistics to attempt to derive a best fit line combines objectivity with some ability to assess the confidence we can have in the relationship. There are two statistical techniques which can be used to provide such a *regression* line of best fit. The simplest of these involves the derivation of the *reduced major axis* regression equation (appendix 5, section 3), initially developed by Pearson in 1901. Earlier it was stated that we could attempt to fit a line to a scatter of points on a graph by minimizing the total deviation between all points and the line. The reduced major axis regression line does this by minimizing the total *area* of the triangles formed between it and all the point values (for example, triangle ABC in figure 3.7). For a résumé of the derivation of the statistical formulae the reader is referred to Kruskal (1953) or Miller and Kahn (1962). The statistic assumes that both sets of data are at least approximately normally distributed (see appendix 5, section 4 and chapter 7). The slope of the regression line is defined by the ratio of the standard deviations of y and x (σ_y and σ_x, see appendix 5, section 1), thus:

$$m = \frac{\sigma_y}{\sigma_x}$$

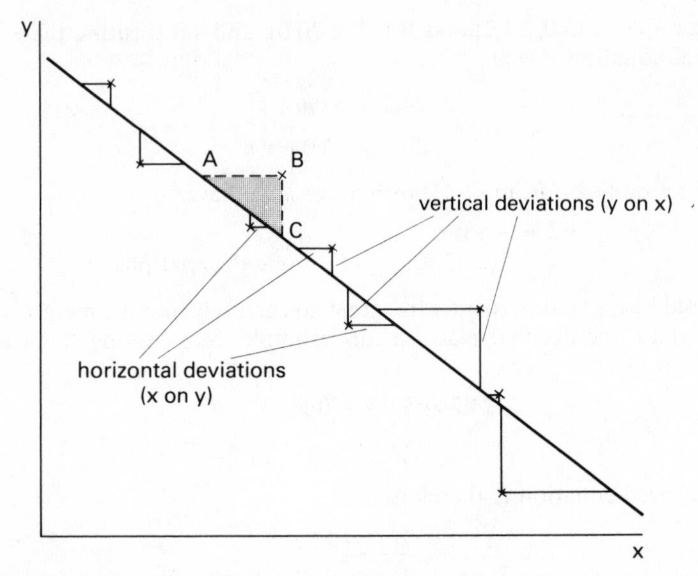

Figure 3.7 Summation of deviations for regression techniques. Minimize areas (ABC) for reduced major axis, horizontal deviations for least-squares regression x on y and vertical deviations for y on x.

and the line also embraces the mean values of both data sets $(\bar{x}, \bar{y})$. The assumption of normality means that certain measures of confidence can be derived for the relationship. For the temperature–altitude relationship shown in figure 3.8, $\bar{t} = 22{\cdot}74$ and $\bar{a} = 8{\cdot}43$, $\sigma_t = 2{\cdot}93$ and $\sigma_a = 5{\cdot}97$. Thus the slope of the regression line is:

$$m = \frac{2{\cdot}93}{5{\cdot}97} = 0{\cdot}49$$

This unfortunately provides no information on the *direction* of the slope of the line, which must first of all be determined with reference to the general trend of data points on a graph. For example where the degree of scatter of points on the graph is high or where the explanation offered by each parameter for the other is poor the technique presents difficulties. In this case however, the slope is clearly negative and therefore substituting in the value $-0{\cdot}49$ for m and the coordinates of the mean values of t and a in the general equation of the straight line we have:

$$22{\cdot}74 = -0{\cdot}49 \times 8{\cdot}43 + k$$
$$k = 26{\cdot}88$$

and the reduced major axis regression line is therefore:

$$\underline{t = 26{\cdot}88 - 0{\cdot}49a} \qquad (3.27)$$

This is shown as line 1 in figure 3.8 and is by coincidence very close to the equation derived from the line drawn in by eye. This would appear to reflect the skill of the drawer of the original line rather than the accuracy of the technique!

Reduced major axis regression provides us with a relatively simple and flexible means of estimating the nature of a relationship from observational

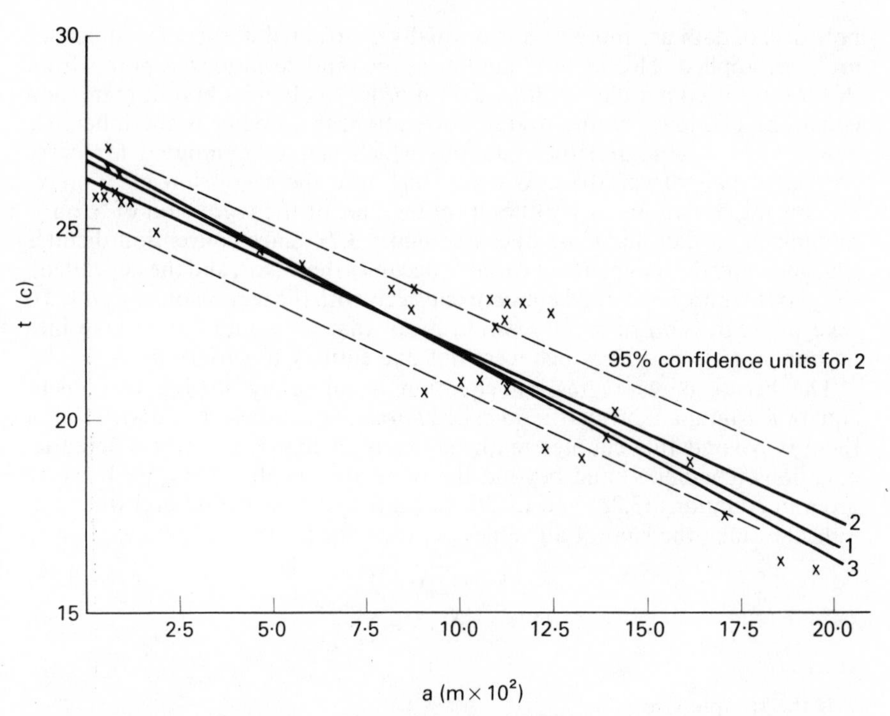

Figure 3.8 Relationship between mean annual temperature *t*, (°C) and altitude a (10^2m) for Tanzania. Line 1, reduced major axis; line 2, *y* on *x* least-squares regression; line 3, *x* on *y* least-squares regression.

data. Whilst it has infinitely more statistical sophistication than the subjective fitting of a line by eye, we are only able to assess the confidence which we can have in the *overall* relationship, and our confidence in predicting *t* from *a* using the equation cannot be objectively assessed. Also the technique makes no assumption as to which variable is dependent and which is independent. The technique is therefore of particular use in areas where independence and dependence cannot be deduced, such as the comparison of morphological variables in palaentology or in the biological sciences. Within physical geography its use is less common than the use of other regression techniques, perhaps because of the search for causality in most physical geography research. However, where causality is not a factor to be established or where it does not have to be demonstrated, the technique provides a useful means of distinguishing between relationships observed within the same population of data. Davidson (1977) has used this very simple technique for this purpose to illustrate different sediment characteristics for upland soils in mid-Wales. Where independence and dependence can be determined, and where confidence in prediction is important, the more rigorous statistical technique of *least-squares regression* should be used. By using this technique we are able to assess our confidence in values of the dependent variable predicted from the equation. Least-squares regression combines objectivity in line-fitting with an assessment of reliability by way of the *product-moment correlation coefficient* (r) (appendix 5, section 2). It is again important to be able to establish that both sets of data are more or less normally distributed if tests of significance are to be applied. This method minimizes the total deviations of points from the line, resolved parallel to *either* axis in order to obtain a best fit regression equation. The lower is the overall correlation, the greater is the difference between the two regression equations which can be computed for every correlated pair of variables. We can thus have the regression of 'y on x', minimizing deviations in y-values from the line, or the regression of 'x on y' minimizing deviations in x-values (see figure 3.7). Since convention dictates that generally the independent variable occupies the x-axis, and the dependent, the y-axis, we are concerned almost exclusively with the regression of y on x. To take the regression of x on y would mean that we would fail to take into account variations in y which were not due entirely to variations in x.

The least-squares regression equation is found by solving two linear equations simultaneously (the so called *normal* equations). The derivation of these is beyond this chapter mathematically in that it involves differential equations (chapter 9) and beyond the book statistically. They are however given in equations (3.28) and (3.29). Using k and m as before and where 'Σ' indicates 'take the sum of all values of', these are:

$$nk + m\sum x = \sum y \tag{3.28}$$

$$k\sum x + m\sum x^2 = \sum xy \tag{3.29}$$

where

n is the sample size
$\sum x$ is the sum of all x-values
$\sum y$ is the sum of all y-values
$\sum x^2$ is the sum of all x^2-values
$\sum xy$ is the sum of all xy product values.

The deduced values for m and k are:

$$m = \frac{n\Sigma xy - \Sigma x \Sigma y}{(n\Sigma x^2 - (\Sigma x)^2)} \tag{3.30}$$

Note here that reference to most statistical texts will reveal that a is used in place of k and b in place of m. We shall continue to use k and m here to maintain continuity.

$$k = \bar{y} - m\bar{x} \tag{3.31}$$

Using the same example as before we have:

$$\bar{x} = \bar{a} = 8{\cdot}43 \quad \text{and} \quad \bar{y} = \bar{t} = 22{\cdot}74,\ n = 35$$
$$x = 295{\cdot}04 \qquad y = 796{\cdot}00$$
$$x^2 = 3735{\cdot}85 \qquad xy = 6136{\cdot}3$$

Substituting in equation (3.30):

$$m = \frac{(35 \times 6136{\cdot}5) - (796{\cdot}0 \times 295{\cdot}04)}{(35 \times 3735{\cdot}9) - (295{\cdot}04 \times 295{\cdot}04)}$$

$$= \frac{-20{,}074{\cdot}3}{130{,}756{\cdot}5 - 87{,}048{\cdot}6}$$

$$= -0{\cdot}46$$

and in (3.31):

$$k = 22{\cdot}74-(-0{\cdot}46\times 8{\cdot}43)$$
$$= 22{\cdot}74+3{\cdot}88$$
$$= 26{\cdot}62$$

The least-squares regression of y on x is therefore:

$$t = 26{\cdot}62-0{\cdot}46a \tag{3.32}$$

and is shown as line 2 in figure 3.8. The calculation of the regression of x on y (line 3 in figure 3.8) reveals:

$$a = 51{\cdot}86-1{\cdot}91t \tag{3.33}$$

or, expressed in a form similar to (3.32):

$$t = 27{\cdot}16-0{\cdot}52a \tag{3.34}$$

Both lines intersect at $(\bar{a}, \bar{t})$, and therefore also intersect with the reduced major axis regression at this point.

Problems 3.5

Plot the data in table 3.1 and in table 3.2 on separate graphs. For each of problems 3.5.1 and 3.5.2 draw in a line by eye and obtain regression equations by both reduced major axis and least-squares regression techniques.

Table 3.1

x	5·0	2·6	1·0	6·8	7·6	9·6	5·6	5·2	6·4	2·0	10·8
y	3·6	2·6	·2·0	5·6	6·4	7·0	5·0	4·4	4·8	2·0	8·2

$\bar{x} = 5{\cdot}53$, $\bar{y} = 4{\cdot}60$, $\sigma_x = 2{\cdot}83$, $\sigma_y = 1{\cdot}89$

Table 3.2

x	33	54	54	5	30	72	43	10	19	45	68	38	30
y	16	35	24	55	35	25	50	40	50	31	28	35	43

$\bar{x} = 38{\cdot}5$, $\bar{y} = 35{\cdot}9$, $\sigma_x = 19{\cdot}7$, $\sigma_y = 11{\cdot}0$

3.6 Significance of results

Although this is not a statistical text, some discussion on the significance of results obtained by regression analysis is in order at this point. More general points are made concerning significance and probability in chapter eight of the companion volume to this (Davidson 1978), or in any general statistical text (e.g. Bryant 1966). It has already been emphasised in chapter 1 and earlier in this chapter that the analysis of very many relationships within physical geography of necessity relies heavily upon the need to take samples. In addition we have already observed that due to errors in instrumentation and the influence of other independent variables, we are in most cases unable to obtain a series of points through all of which a straight line may be drawn on a graph. In the previous example the points plotted on the graph in figure 3.8

approximated a straight line relatively closely. Because of this the two least-squares regression lines were very close and their equations similar. The reduced major axis line bisects these two. If we were to use the derived function relating temperature to altitude for predictive purposes we should be able to predict with a certain confidence the temperature for a given altitude. Now consider the situation in figure 3.9 where another set of data pairs have been plotted on a graph. In this example a possible relationship between temperature contrast with the arrival of a sea breeze at Lampeter, Wales (T_c), and the original contrast between air mass and sea surface termperature (T_s) is being tested (Sumner 1977b). There is a proportionately great scatter of points over the graph to the extent that it is difficult to draw a line in by eye. The least-squares regression lines have been calculated and plotted, for which the equations are:

T_c on T_s $\qquad T_c = 0{\cdot}08T_s + 0{\cdot}65 \qquad (3.34)$

T_s on T_c $\qquad T_c = 0{\cdot}80T_s - 2{\cdot}68 \qquad (3.35)$

There is a marked difference between the two lines, and any attempt at using these for prediction would yield a result which we could not expect to bear much resemblance to reality.

Thus, in addition to calculating the regression equations themselves we require some means of expressing our faith in the results, or the statistical significance of the relationship. The key to this lies with the calculation of the correlation coefficient. The value of the coefficient lies always between $-1{\cdot}0$ and $+1{\cdot}0$, with a value of $+1{\cdot}0$ indicating perfect correlation with increasing x being associated with increasing y. When plotted on a graph all points would exactly coincide with the regression equation line, and both regression lines will coincide. The slope of the regression line is therefore exactly the ratio between the two standard deviations, σ_x and σ_y. A correlation of $-1{\cdot}0$ would indicate a similar perfect relationship but with increasing y associated with *decreasing* x, and correspondingly, a negative slope, but whose modulus is still the ratio of the standard deviations. For both positive and negative correlations, as the magnitude of $|r|$ decreases so the regression lines diverge, until at $r = 0$ the two are mutually at right-angles. The correlation coefficient

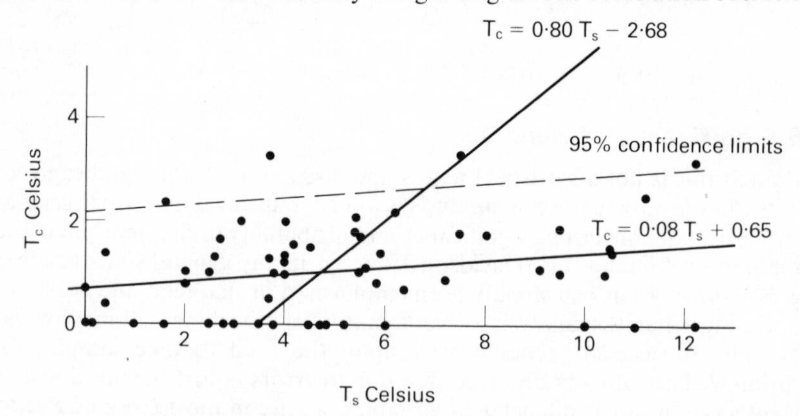

Figure 3.9 Relationship between temperature contrast with arrival of a sea breeze (T_c) and air mass/sea temperature contrast (T_s), Lampeter, Wales, 1973–1975.

for the data used in figure 3.8 was $-0{\cdot}9372$, whilst that between T_c and T_s in the current example is only 0·3108. The degree of statistical explanation of the dependent variable offered by the independent is given by r^2, so that in our first example we can account for about 88% of the variation of t by variation in a, but in the present example, only 10% of variation in T_c is explained by variation in T_s.

In addition, the confidence we can have in the correlation also depends upon the size of the sample. Quite simply, even though we may have a high correlation coefficient, if our sample is very small then there is a relatively greater chance of spurious readings unduly influencing the form of the regression equation. The significance of the correlation coefficent therefore is generally expressed in terms of the probability that the derived relationship using the sample data would be true for the total population. This significance can be arrived at using the *Student's t* test (appendix 5, section 5) and tables. The t-tables assume that data are normally distributed, underlining again the need to ensure that both data sets are close to being normally distributed from the outset. The result of applying such a test to a correlation coefficient of $-0{\cdot}9372$ for a sample size of $n = 35$ reveals that the relationship between temperature and altitude in Tanzania is significant to less than the 0·1% rejection level, or that we can be more than 99·9% confident that our regression line portrays a relationship which is true for all locations in the whole of the country. In the second example we obtain a value for Student's t of 2·5, which is significant only to the 2% rejection level for $n = 61$, so that we can be only 98% confident in this case.

If we are aiming to use the regression equations for predictive purposes then we must use the correlation coefficient in conjunction with the standard deviation of the dependent variable to compute the *standard error* (SE) of y where:

$$\text{SE} = \sigma_y\sqrt{(1-r^2)} \tag{3.36}$$

Assuming again that data are normally distributed then approximately 95% of values of y predicted from the regression will lie between $-2\text{SE} \leq y \leq +2\text{SE}$. In the first example the standard error of y is 0·51, so that for example, at a height of 500 metres the predicted mean annual temperature from equation (3.32) would be 24·3 °C, but that on 95% of occasions we should find the temperature at that altitude between 24·3 − 1·02 °C and 24·3 + 1·02 °C: 23·3 to 25·3 °C. By drawing in two lines on the graph parallel to that of equation (3.32), but at $t = -1{\cdot}02$ and $t = +1{\cdot}02$ away from it, we shall have constructed the 95% confidence limits of the regression equation. A similar treatment of the second example reveals a standard error of 0·76, so that for $T_s = 5$ °C we have from equation (3.34), $T_c = 1{\cdot}05$ °C. With 95% confidence a temperature contrast of 5·0 degrees Celsius between the temperature of the sea surface and the overlying air mass will be associated with a temperature change of 1·1 °C ±1·5 °C on the arrival of a sea breeze; a very poor predictor indeed!

It has been impossible in the very short space allowed to give a very detailed résumé of correlation and significance levels. Such topics belong in the realm of statistics and the reader is referred again to a good elementary statistics text, such as Gregory (1970) or Bryant (1966), for a more detailed explanation. However the point must be made that it is with the ability for an objective statistical appraisal of confidence in predicted results that the least-squares technique scores over other methods. With the reduced major axis technique it is not possible to produce standard errors of the estimate for predicted values of y. Only the standard error of the slope and the intercept can be calculated. No statistical significance can be ascribed to equations derived purely from lines drawn by eye. The reader will now find it useful and instructive to relate the correlation coefficients and computed standard errors for the data given in tables 3.1 and 3.2 (problems 3.5.1 and 3.5.2) to the regression equations he has calculated and plotted for these exercises. Product-moment correlation coefficients for the data are respectively 0·9853 and −0·6183.

3.7 Multivariate relationships

The treatment of bivariate relationships has served to illustrate the basis of linear functions. The relatively simple portrayal of such functions considerably eases understanding of functions as a whole. Physical geography, however, exists in a multivariate world and it is comparatively rare to obtain a complete explanation of one variable solely in terms of one other, even assuming there are no measurement errors. It is often possible however to explain *most* of the variation in the dependent variable in terms of one independent variable. In the example of temperature and altitude in Tanzania the total statistical explanation offered was 88%. It is important to stress the difference between statistical and scientific explanation. 88% statistical explanation does not necessarily infer 88% scientific explanation. Error factors, totally independent of physical geographical influences, will tend to decrease the amount of apparent explanation. This of course, is not to say that we should not still search for other geographical influences in order to increase the level of overall explanation, and multivariate statistics exist to permit the derivation of statistical models along similar lines to bivariate correlation and regression. The computation, although not the mathematics, involved is naturally more complex, frequently demanding computer analysis, and for this reason multivariate functions are not dealt with in as much detail as more simple functions. Students wishing to follow up the statistics behind multivariate relationships should consult advanced texts, such as Ezekiel and Fox (1959), Bryant (1966), or for more specialist treatment in aspects of physical geography, Miller and Kahn (1962) and Panofsky and Brier (1958).

Multivariate functions at their most elementary involve three variables. The extension of techniques developed for bivariate relationships is relatively simple and is brought about by the addition of another graph axis, the z-axis. The resulting set of axes is thus in three dimensions, but this can be coped with without too much confusion on two-dimensional paper (figure 3.10). Any point can be defined with reference to these axes by a set of three coordinates. The graphical representation of a function containing three variables is as a *surface*, rather than the line of functions involving only two dimensions. The computation of trend surfaces, for example, those computed for a rock platform by Gray (1974), fits mathematical functions in three dimensions as approximations to the spatial scatter of data points. A full understanding of multiple regression and correlation analysis demands that the student is conversant with the concepts and operations of matrix algebra (chapter 4). This is particularly important when attempting to construct mathematical models involving more than three variables. Since we find it difficult to

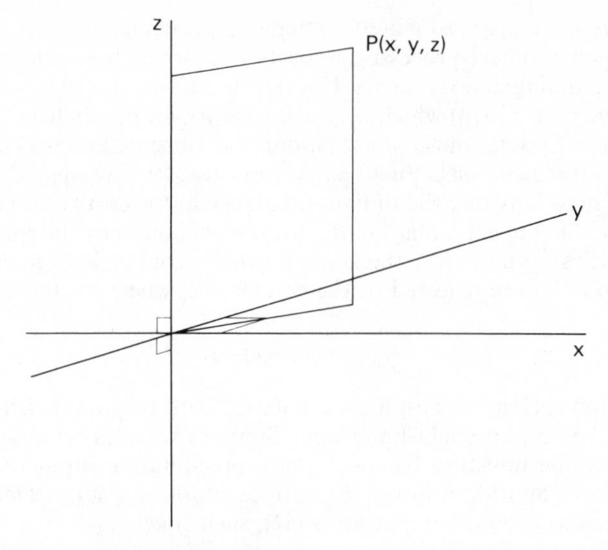

Figure 3.10 Three-dimensional axes.

comprehend three-dimensional space though, it is equally difficult to represent such functions graphically without confusion arising, and we are thus totally dependent upon mathematical notation. Matrix algebra becomes an important part of such notation.

An example of a very basic function involving a number of independent variables was given in chapter 1, when it was proposed that soil type (s) was determined by a number of independent factors, climate (c), vegetation (v), organisms (o), parent material (p), rainfall (r) and time (t) in:

$$s = f(c,v,o,p,r,t) \tag{3.37}$$

Multivariate statistical techniques would have to be used in order to determine the mathematical function from field data. A more simple model, this time relating storm rainfall (P_s) and antecedent precipitation (API) to storm runoff in a river (Q_s) is given by Weyman (1975) for a site in Somerset:

$$Q_s = -3{\cdot}16 + 0{\cdot}35P_s + 0{\cdot}34\text{API}$$

This could be represented as a linear surface with reference to three graph axes. Such linear multivariate functions are relatively rare in physical geography, and even in this example the derivation of the antecedent precipitation index involves the weighting of previous daily precipitation totals according to a power function (see chapters 5 and 6). Both P_s and API contribute to the variation in Q_s and overall the expression provided a 65% statistical explanation of the total variation in storm runoff. If it had been possible to include additional independent variables the explanation would almost certainly have been increased.

3.8 Continuity equations

There are many examples of functions involving more than two variables in physical geography whose scientific explanation is by definition complete.

Frequently these are used merely to depict a particular system. The study of physical geographical processes as parts of systems has widened in recent years, culminating in texts such as Chorley and Kennedy (1971) or Russwurm and Sommerville (1974) which approach the topics purely from the systems point of view. Systems must balance inputs and outputs, and may be described by *continuity equations.* Such equations are for example developed in climatology to illustrate the radiation balance at the earth's surface. Thus in equation (3.38) the net radiation (R_n) must be balanced by the solar radiation at the earth's surface (R_c), the outgoing terrestrial radiation (R_b) and the amount of radiation reflected by the surface (rR_c where r is the albedo of the surface):

$$R_n = R_c(1-r) - R_b \tag{3.38}$$

By definition such an equation must balance, with the inputs (left-hand side) equalling the outputs (right-hand side). Similar examples occur in hydrology to illustrate the moisture balance, where precipitation input (P) is exactly balanced by evaporation losses (E), surface runoff (R), subsurface runoff (S) and the loss to groundwater storage (M), such that:

$$P = E + R + S + M \tag{3.39}$$

The same general form of equation can be utilized when considering soil moisture conditions for much smaller volumes. Such an equation can be used to estimate a parameter whose magnitude cannot be directly measured by instrumentation. For example it is relatively easy to measure P and E within a catchment, albeit with limited precision, and to monitor S and R by their contribution to river discharge at the exit from the catchment. In doing so an assessment of the change in stored moisture over a period of time, can be made. We are in other words, attempting to determine the *rate* of addition of water to the water table. Recourse to the theory of water movement through the soil under changing moisture conditions (see Hudson 1971) reveals however that this rate of movement and therefore the rate of addition of moisture to the water table, is dependent in turn upon the quantity of water already present (see, for example, Quinn 1977). A further factor is thus introduced which will strongly influence the magnitude of the other variables relative to one another. The rate of addition of moisture to storage will not be constant for constant rainfall rate. For example, as the soil becomes saturated and its infiltration capacity is reached, far more water will remain on the surface for removal by evaporation and surface runoff, and thus while M decreases, R and E increase to compensate and maintain the balance. Such an equation therefore can be more complex in its behaviour than its apparent simplicity would suggest, and its detailed solution for a particular set of circumstances frequently demands recourse to differential equations (chapter 9).

Many bivariate relationships within physical geography are however non-linear in character, and similarly many multivariate relationships are also non-linear. Non-linear ones tend to be the rule and linear ones the exception where models are constructed based upon field or laboratory observation. The fitting of curves to data points rather than straight lines is complicated only in the sense that the *form* of the relationship which most approaches the observed scatter of points has first to be established. For example, many involve logarithmic functions, as in:

$$y = 32{\cdot}2 \log x + 0{\cdot}8 \tag{3.40}$$

Substitution here of $t = \log x$ reveals an equation of a form identical to the standard equation for a linear function. This provides us with a clue as to how to approach the problem in chapter 5. However, before we leave linear relationships entirely there is one development of the regression models established in section 3.5, principal components analysis, which is best approached once we have introduced matrix algebra in the next chapter.

4

Matrices

4.1 The basic matrix

Matrices, like vectors are now commonly used within physical geography, and are particularly useful when undertaking advanced statistical analysis of data, as we shall see in section 4.10 when matrix algebra is applied to the finding of principal components. The most elementary types of matrix are in fact vectors. However it is usual to depict matrices not in vector form but in table form. The most simple example of a matrix is called either a *row* or a *column* matrix depending upon whether the sequence of values is respectively horizontal or vertical. A single string of numbers constitutes a row or column matrix. Since direction is implied by the terms themselves, alternative names for these types of matrix are row and column vectors. The sequence of values indicating mean monthly temperature as normally presented in text books constitutes a row matrix. The simple listing of values down a page would similarly constitute a column matrix.

More commonly however, matrices consist of a number of rows and columns so that the overall impression is of a table of values. As an illustration some data are portrayed in table 4.1 in conventional tabular form as they

Table 4.1 Choice of physical or human geography by men and women students.

	Men	Women	Total
Physical	20	40	60
Human	33	15	48
Regional	7	4	11
Total	60	59	119

might appear in any text. The table shows the stated preference of men and women geographers at a university department for physical, human and regional geography. It will be seen that overall physical geography is far more popular than either human or regional geography, and that a much higher proportion of women students prefer it (68%) than do men students (33%)—this of course is quite normal! Excluding the total from each column and row we have a 3×2 matrix (3 rows and 2 columns) which can be written:

$$A = \begin{pmatrix} 20 & 40 \\ 33 & 15 \\ 7 & 4 \end{pmatrix}$$

where A is the name of the matrix. The individual compartments in the matrix are called *elements, entries* or *components* and the general expression for a matrix A_{mn} of m rows and n columns is:

$$A_{mn} = \begin{pmatrix} i_{11} & i_{12} & i_{13} & i_{14} & i_{15} & \cdots & i_{1n} \\ i_{21} & i_{22} & i_{23} & i_{24} & i_{25} & \cdots & i_{2n} \\ i_{31} & i_{32} & i_{33} & i_{34} & i_{35} & \cdots & i_{3n} \\ i_{41} & i_{42} & i_{43} & i_{44} & i_{45} & \cdots & i_{4n} \\ i_{51} & i_{52} & i_{53} & i_{54} & i_{55} & \cdots & i_{5n} \\ \vdots & \vdots & \vdots & \vdots & \vdots & & \vdots \\ i_{m1} & i_{m2} & i_{m3} & i_{m4} & i_{m5} & \cdots & i_{mn} \end{pmatrix} \tag{4.1}$$

This is a matrix of dimension mn. In the previous example the dimensions were $2 \times 3 = 6$. Those readers who may already have had experience of computer programming will perhaps have noticed the close similarity between a matrix A_{mn} and an *array* of dimension mn, or in Fortran, $A(m,n)$. If $m = n$ then the matrix shown in equation (4.1) is said to have a *diagonal* composed of the elements i_{11}, i_{22}, i_{33}, i_{44} and so on. Such a matrix where $m = n$ is called, for reasons which should be obvious, a *square* matrix.

The nature of the variation in element values through any matrix determines whether or not the matrix is *symmetric.* For example, in most gazeteers there appears a tabulation of the distances apart of the major towns and cities. If the complete matrix were shown then the part below the diagonal will mirror exactly that above the diagonal, since the sequence of towns will be the same along both the rows and columns. A more simple, abstract, example is shown here:

$$\begin{pmatrix} 1 & 2 & 0 \\ 2 & 2 & 5 \\ 0 & 5 & 3 \end{pmatrix}$$

is clearly symmetric since there is a mirror-image of element values about the diagonal, whilst,

$$\begin{pmatrix} 1 & 2 & 0 \\ -1 & 2 & 5 \\ 0 & 6 & 3 \end{pmatrix}$$

is not symmetric. A matrix which is not symmetric is often termed *skew-symmetric.* Again, by definition, a symmetric matrix must also be a square matrix. One very important type of symmetric matrix is the *unit* or *identity matrix* (I), where the values of elements along the diagonal are all equal to one, whilst all other elements are zero. This has particularly useful properties which will be used in later sections of this chapter. For example, we shall see that the unit matrix acts as if it were a scalar number one in matrix multiplication, so that:

$$\begin{pmatrix} 1 & 0 & 0 \\ 0 & 1 & 0 \\ 0 & 0 & 1 \end{pmatrix} \begin{pmatrix} 1 \\ 2 \\ 3 \end{pmatrix} = \begin{pmatrix} 1 \\ 2 \\ 3 \end{pmatrix} \tag{4.2}$$

In addition a matrix where all the elements have the value zero is called a *null matrix*:

$$\begin{pmatrix} 0 & 0 & 0 \\ 0 & 0 & 0 \\ 0 & 0 & 0 \end{pmatrix} \tag{4.3}$$

4.2 Equality of matrices

If we have two matrices A and B where:

$$A = \begin{pmatrix} 0 & 4 \\ a & b \end{pmatrix} \quad \text{and} \quad B = \begin{pmatrix} c & 4 \\ 2 & 0 \end{pmatrix}$$

then $A = B$ if and only if (iff) $b = c = 0$ and $a = 2$. Similarly if we let two matrices A and B take on the following elements:

$$A = \begin{pmatrix} x-2 \\ 5y+2 \end{pmatrix} \quad B = \begin{pmatrix} y+1 \\ 4+x \end{pmatrix}$$

then $A = B$ iff:

$$x-2 = y+1 \tag{4.4}$$

and:

$$4+x = 5y+2 \tag{4.5}$$

subtracting (4.5) from (4.4) we have:

$$-6 = -4y-1$$

or

$$y = 1\tfrac{1}{4}$$

Substituting this into (4.4) we have:

$$x-2 = 1\tfrac{1}{4}+1$$

or

$$x = 4\tfrac{1}{4}$$

Clearly then we are able to represent and solve, simultaneous equations in matrix form (see also section 3.2). This will be developed further when we have considered the operations of matrix addition and multiplication. Consider now the situation where:

$$A = \begin{pmatrix} x+y & 5 \\ 3+2y & y \end{pmatrix} \quad B = \begin{pmatrix} 6 & y \\ 3x+1 & x+y \end{pmatrix}$$

then for row one:

$$x+y = 6 \text{ and } y = 5, \text{ so that } x = 1 \text{ and } y = 5$$

but for row two:

$$3+2y = 3x+1 \text{ and } y = x+y, \text{ whence } x = 0 \text{ and } y = -1$$

Thus matrix A cannot be equal to matrix B.

4.3 Computation with matrices

Most arithmetic operations with matrices follow on logically from similar computations concerning scalar quantities, but, as with vector multiplication, an exception is to be found in matrix multiplication. Let us first of all illustrate the simple case of matrix addition. Consider the case where the mean occurrence of days with rain is known for a site on a seasonal basis for autumn, winter and spring, and that these are 64, 57 and 46 days respectively. We may represent this data in the form of a row vector A, where:

$$A = (64 \quad 57 \quad 46)$$

a matrix of 1×3 elements. The sum of the element values within the matrix gives the row sum representing the average number of rain days occurring in all three seasons taken together. If we wished to estimate the probable number of such days on five successive years we should simply find the product of the matrix A and the scalar five, such that:

$$5A = (320 \quad 285 \quad 230)$$

Clearly then the product of a scalar and a matrix is a simple operation of multiplying each element in the matrix by the scalar value to produce the new matrix.

If we now suppose that similar data are available for the daily occurrence of snowfall at the same location and turn our attentions to a consideration of the nature of *specific* seasons as reflected by snow and rain incidence, we can construct two 2×3 matrices to illustrate the operations involved with matrix addition. If in year x there were respectively 70, 64 and 39 days with rain, and 4, 10 and 8 days with snow for the three seasons, and comparable data for year y were 71, 38 and 32 days with rain, and 0, 35 and 10 days with snow, then by defining matrix A as expressing rain incidence and a matrix B expressing snow incidence in each year, we may write:

$$A = \begin{pmatrix} 70 & 64 & 39 \\ 71 & 38 & 32 \end{pmatrix} \quad \text{and} \quad B = \begin{pmatrix} 4 & 10 & 8 \\ 0 & 35 & 10 \end{pmatrix}$$

In each matrix the rows represent data for each *year* and the columns give data for *seasonal incidence.* Year x (top row) was clearly relatively wet and mild, whilst year y was particularly snowy in the winter season. The addition of the two matrices to produce a third matrix C, will yield the combined incidence of rain and snow in each year, and its derivation is a simple matter of adding together each of the individual elements in each array, such that:

$$\begin{pmatrix} 70 & 64 & 39 \\ 71 & 38 & 32 \end{pmatrix} + \begin{pmatrix} 4 & 10 & 8 \\ 0 & 35 & 10 \end{pmatrix} = \begin{pmatrix} 74 & 74 & 47 \\ 71 & 73 & 42 \end{pmatrix}$$

Note that although this operation is a very simple extension of scalar arithmetic, we are only able to add together matrices which have the *same number of elements along each row and down each column.*

To illustrate the law of matrix multiplication let us now apply the data to a simple economic problem. A local transport company in the area for which the above data apply is able to cost the effect of rain, snow and fog incidence against delays. If we combine the occurrence of each of these for the two years regardless of season we can construct a 2×3 matrix in which the third column indicates fog incidence and the first two, the incidence of snow and rain respectively, so that $C_{11} = B_{11} + B_{12} + B_{13}$ and $C_{12} = A_{11} + A_{12} + A_{13}$ and so on. This new matrix C is therefore:

$$C = \begin{pmatrix} 22 & 173 & 12 \\ 45 & 141 & 15 \end{pmatrix}$$

Let us now indicate the cost per day of delays induced by snow (D_{11}), rain (D_{21}) and fog (D_{31}) respectively, as a column matrix, D:

$$D = \begin{pmatrix} 50 \\ 10 \\ 20 \end{pmatrix}$$

In order that we may multiply matrices together we must have the same number of columns in the second matrix as we have rows in the first. It is of course equally important to ensure that the variable represented by the columns of the first and the rows of the second relate to the same items! To multiply matrices C and D it should be clear that for each element in the first row of C we must multiply by the appropriate element in matrix D. The resulting matrix, CD, is therefore given by:

$$CD = \begin{pmatrix} 22 & 173 & 12 \\ 45 & 141 & 15 \end{pmatrix} \begin{pmatrix} 50 \\ 10 \\ 20 \end{pmatrix} = \begin{pmatrix} 22 \times 50 + 173 \times 10 + 12 \times 20 \\ 45 \times 50 + 141 \times 10 + 15 \times 20 \end{pmatrix}$$

$$= \begin{pmatrix} 3070 \\ 3960 \end{pmatrix}$$

The overall cost of year x was thus 3070 units and that of year y was 3960 units. The same rules of multiplication are extended to matrices of any dimension, with the proviso that for any two matrices A_{mn} and B_{pq}, p must equal n. In the above example matrix C had dimensions 2×3, whilst matrix D had dimensions 3×1. The operation yielded a matrix of dimension 2×1 and thus:

$$C_{mn}\, D_{pq} = E_{mq} \tag{4.6}$$

To illustrate the case where the second matrix has more than one column we can add a further column vector to matrix D (say, reflecting the cost incurred through *accidents* under conditions of snow, rain and fog). We now have to solve:

$$CD = \begin{pmatrix} 22 & 173 & 12 \\ 45 & 141 & 15 \end{pmatrix} \begin{pmatrix} 50 & 10 \\ 10 & 1 \\ 20 & 8 \end{pmatrix}$$

$$= \begin{pmatrix} 22 \times 50 + 173 \times 10 + 12 \times 20 & 22 \times 10 + 173 \times 1 + 12 \times 8 \\ 45 \times 50 + 141 \times 50 + 15 \times 20 & 45 \times 10 + 141 \times 1 + 15 \times 8 \end{pmatrix}$$

$$= \begin{pmatrix} 3070 & 489 \\ 3960 & 711 \end{pmatrix}$$

Again we see that equation (4.6) holds, and that the product matrix has as many *rows* as the *first* matrix, and as many *columns* as the *second* matrix. Note that the order of computation is *row first* and *column second*. Thus matrix products are non-commutative, since the product DC from above must yield a matrix with dimension 3×3. Hence in general terms for matrices:

$$A.B \neq B.A \tag{4.7}$$

Problems 4.3

Obtain $A.B$ where:

4.3.1 $$A = \begin{pmatrix} 3 & 2 & 5 \\ 2 & 8 & 0 \end{pmatrix} \quad B = \begin{pmatrix} 2 & 4 \\ 3 & 1 \\ 0 & 1 \end{pmatrix}$$

4.3.2 $$A = \begin{pmatrix} -2 & 4 & 6 \\ 3 & 4 & 10 \\ 2 & 0 & -5 \end{pmatrix} \quad B = \begin{pmatrix} 17 \\ -5 \\ 2 \end{pmatrix}$$

4.3.3 Determine values for x and y if matrix A = matrix B

$$A = \begin{pmatrix} 3y+2x & 10x-y \\ 3x-2y & 9y+x \end{pmatrix} \quad B = \begin{pmatrix} 3x+y & 8x-15 \\ 2x & 5x+y \end{pmatrix}$$

4.4 The laws of matrix algebra

At this stage, before we advance to further topics concerning matrices it is as well to summarize the laws of matrix algebra so far inferred or established.

4.4.1 Associative Law for addition and multiplication

$$A+(B+C) = (A+B)+C$$
$$A(BC) = (AB)C$$

4.4.2 Commutative Law (for addition only)

$$A+B = B+A$$

4.4.3 Distributive Law for addition and multiplication

$$A(B+C) = AB+AC$$
$$k(A+B) = kA+kB$$
$$k(AB) = kAB$$
$$(k+l)A = kA+lA$$
$$(kl)A = k(lA)$$

With the exception of the commutative law which holds only for addition, the laws are the same as for scalar algebra.

4.5 Transposition of matrices

A transposed matrix is one in which the column and row elements have been exchanged. Thus the transpose of a column matrix will be a row matrix and vice versa. In general terms if a matrix A has the form:

$$A = \begin{pmatrix} i_{11} & i_{12} \\ i_{21} & i_{22} \\ i_{31} & i_{32} \end{pmatrix}$$

then its transpose, A' or A^{T} is:

$$A^{\mathrm{T}} = \begin{pmatrix} i_{11} & i_{21} & i_{31} \\ i_{12} & i_{22} & i_{32} \end{pmatrix}$$

A further property is that the product of two matrices A and B is equalled by the product of the transpose forms of those matrices, B and A, or:

$$AB = B^{\mathrm{T}}A^{\mathrm{T}} \tag{4.8}$$

We can demonstrate this as follows: let

$$A = \begin{pmatrix} i_{11} & i_{21} \\ i_{12} & i_{22} \end{pmatrix} \quad \text{and} \quad B = \begin{pmatrix} j_{11} \\ j_{21} \end{pmatrix}$$

so that:

$$A^{\mathrm{T}} = \begin{pmatrix} i_{11} & i_{21} \\ i_{12} & i_{22} \end{pmatrix} \quad \text{and} \quad B^{\mathrm{T}} = (j_{11} \quad j_{21})$$

and thus:

$$\begin{aligned} AB &= (i_{11}j_{11}+i_{12}j_{21} \quad i_{21}j_{11}+i_{22}j_{21}) \\ B^{\mathrm{T}}A^{\mathrm{T}} &= (j_{11}i_{11}+j_{21}i_{12} \quad j_{11}i_{21}+j_{21}i_{22}) \end{aligned}$$

4.6 The inverse matrix

In ordinary numeric operations we refer to the reciprocal of a number and can define it as $k\frac{1}{k} = 1$, where the number $\frac{1}{k}$ is the reciprocal of k. The equivalent in matrix terms is known as the inverse matrix, such that:

$$AB.BA = I \quad \text{(identity matrix)} \tag{4.9}$$

where B is the inverse of A. However the operation involved in determining the inverse of A is not as simple as its equivalent is for scalar operations. If we take a matrix A, where:

$$A = \begin{pmatrix} 2 & 4 \\ 6 & 8 \end{pmatrix} \quad \text{and let} \quad B = \begin{pmatrix} a & b \\ c & d \end{pmatrix}$$

then we know that:

$$\begin{pmatrix} 2 & 4 \\ 6 & 8 \end{pmatrix}\begin{pmatrix} a & b \\ c & d \end{pmatrix} = \begin{pmatrix} 1 & 0 \\ 0 & 1 \end{pmatrix}$$

from our definition (4.9), so that from section 4.3 we can multiply these out to give four simultaneous equations:

$$2a+4c = 1 \tag{4.10}$$

$$2b+4d = 0 \tag{4.11}$$

$$6a+8c = 0 \tag{4.12}$$

$$6b+8d = 1 \tag{4.13}$$

Solving (4.12) and (4.10) simultaneously we obtain:

$$b = \tfrac{1}{2} \quad \text{and} \quad d = -\tfrac{1}{4}$$

so that:

$$B = \begin{pmatrix} -1 & \frac{1}{2} \\ \frac{3}{4} & -\frac{1}{4} \end{pmatrix} \text{ is the inverse of } A = \begin{pmatrix} 2 & 4 \\ 6 & 8 \end{pmatrix}$$

and we can write:

$$B = A^{-1} \tag{4.14}$$

In a general form for this arrangement of matrices we thus have:

$$\begin{pmatrix} p & q \\ r & s \end{pmatrix}\begin{pmatrix} a & b \\ c & d \end{pmatrix} = \begin{pmatrix} 1 & 0 \\ 0 & 1 \end{pmatrix}$$

so that:

$$ap + qc = 1 \tag{4.15}$$

$$bp + qd = 0 \tag{4.16}$$

$$ra + sc = 0 \tag{4.17}$$

$$rb + sd = 1 \tag{4.18}$$

and from these:

$$p = d/(ad - bc) \tag{4.19}$$

$$q = -b/(ad - bc) \tag{4.20}$$

$$r = -c/(ad - bc) \tag{4.21}$$

$$s = a/(ad - bc) \tag{4.22}$$

Thus the inverse of the matrix $\begin{pmatrix} a & b \\ c & d \end{pmatrix}$ is:

$$\frac{1}{ad - bc}\begin{pmatrix} d & -b \\ -c & a \end{pmatrix} \tag{4.23}$$

The expression $(ad - bc)$ is the *determinant* of the matrix, and can be written:

$$\det A \quad \text{or} \quad \det \begin{pmatrix} a & b \\ c & d \end{pmatrix} \quad \text{or} \quad \begin{vmatrix} a & b \\ c & d \end{vmatrix} \tag{4.24}$$

Determinants may be found, though less easily, for square matrices of any dimension, as we shall see in section 4.7. Some matrices however, have no inverse form. In such circumstances the determinant has the value zero. For example, the matrix:

$$\begin{pmatrix} 2 & 6 \\ 4 & 12 \end{pmatrix}$$

has no inverse since its determinant $(ad–bc)$ is zero. Indeed all matrices of the form:

$$\begin{pmatrix} a & b \\ ka & kb \end{pmatrix}$$

have no inverse. Such matrices are referred to as *singular matrices*, whilst by definition, those matrices for which an inverse can be obtained are *non-singular*.

4.7 Determinants

Determinants are particularly useful if we are to use matrix algebra to solve simultaneous equations (section 4.9), although for matrices whose dimensions exceed 2×2 the task is rather more difficult than that shown above. Let us now consider the case where we must determine the third-order determinant—the determinant of a 3×3 matrix—using the notation which appears in equation (4.1). We could, if we wished, arrive at a further general solution for the determinant, using a process similar to that developed in equations (4.15) to (4.23). However, this becomes rather laborious, and we may define the determinant for *any* dimension square matrix in relatively simple terms.

For any matrix A_{mn}, the determinant may be defined as:

$$|A| = \sum_{i=1}^{i=m} (-1)^{i+j} a_{ij} \alpha_{ij} \tag{4.25}$$

where a_{ij} is the value of each element in row i, column j, and α_{ij} is known as the 'minor' of the element a_{ij}. The minor may be defined in terms of the *cofactor* (A_{ij}) of the element a_{ij}, or:

$$A_{ij} = (-1)^{i+j} \alpha_{ij} \tag{4.26}$$

Using the notation from equation (4.1) the derivation of a minor for the elements a_{11} and a_{21} in a 3×3 matrix is as indicated in equation (4.27):

$$\begin{array}{c|cc} \textcircled{a_{11}} & a_{12} & a_{13} \\ \hline \textcircled{a_{21}} & a_{22} & a_{23} \\ \hline a_{31} & a_{32} & a_{33} \end{array} \tag{4.27}$$

Thus,

$$\alpha_{11} = \begin{vmatrix} a_{22} & a_{23} \\ a_{32} & a_{33} \end{vmatrix} \tag{4.28}$$

$$\alpha_{21} = \begin{vmatrix} a_{12} & a_{13} \\ a_{32} & a_{33} \end{vmatrix} \tag{4.29}$$

$$\alpha_{23} = \begin{vmatrix} a_{11} & a_{12} \\ a_{31} & a_{32} \end{vmatrix} \tag{4.30}$$

and so on. At this stage it is important to note the positions of elements included in these matrices. In each case the row and column (i and j) in which a_{ij} is placed are omitted from consideration, and the elements in the four corners of the remainder of matrix A_{mn} comprises the new matrix of the minor α_{ij}. Effectively we are reducing the larger matrix to factors of a series of smaller matrices for which the calculation of the determinant is a relatively straightforward process. For matrices of larger dimensions than 3×3 we should have to proceed stage by stage through a series of minor 2×2 matrices. The determinants of the minors (α_{ij}) are the cofactors of the elements a_{ij}. Having found these cofactors of each element a_{ij} the solution for the determinant of the larger matrix is relatively easy, but before proceeding with a worked example two important points must be emphasized. These will simplify the operations and clarify the definition given in equation (4.25). Firstly, the sign of the cofactors of A_{ij}, determined in equation (4.26) by the term $(-1)^{i+j}$, merely alternates for alternate element locations. Thus the sign of α_{11} is $+$, that of α_{12} is $-$, α_{13} $+$, α_{21} $-$, and so on. We can accordingly produce a matrix of signs of the cofactors A_{ij}, which for a 3×3 matrix is:

$$\begin{pmatrix} + & - & + \\ - & + & - \\ + & - & + \end{pmatrix} \tag{4.31}$$

The second point is that we may calculate the determinant using only elements from *one* column (a_{ij}, where j is constant) and calculating α_{ij} from the remaining rows and columns (see equations (4.27) to (4.30)). The computation of the determinant value using any or all other columns should be the same, assuming our arithmetic is correct!

Let us now take the matrix A defined as:

$$A = \begin{pmatrix} 3 & 2 & 1 \\ 2 & 1 & -4 \\ 1 & -1 & -1 \end{pmatrix} \tag{4.32}$$

Taking the elements of the first column and substituting into equation (4.25) we have:

$$\begin{aligned} |A| &= 3\begin{vmatrix} 1 & -4 \\ -1 & -1 \end{vmatrix} - 2\begin{vmatrix} 2 & 1 \\ -1 & -1 \end{vmatrix} + 1\begin{vmatrix} 2 & 1 \\ 1 & -4 \end{vmatrix} \\ &= 3 \times (-1-5) - 2 \times (-2+1) + 1 \times (-8-1) \\ &= -22 \end{aligned}$$

Repeating the exercise using column two we have:

$$\begin{aligned} |A| &= -2\begin{vmatrix} 2 & -4 \\ 1 & -1 \end{vmatrix} + 1\begin{vmatrix} 3 & 1 \\ 1 & -1 \end{vmatrix} + 1\begin{vmatrix} 3 & 1 \\ 2 & -4 \end{vmatrix} \\ &= -2 \times (-2+4) + 1 \times (-3-1) + 1 \times (-12-2) \\ &= -22 \quad \text{as before.} \end{aligned}$$

4.8 The adjoint matrix

In order that we may now apply the determinant values to matrices to find their inverse forms we must first of all define the *adjoint* matrix: Adj A. The adjoint matrix is simply the transpose of the cofactors of all the elements a_{ij} in a matrix A_{mn}. Thus for the matrix A:

$$A = \begin{pmatrix} 5 & 2 & 4 \\ 3 & 1 & 6 \\ 0 & 1 & 2 \end{pmatrix}$$

the matrix of cofactors of A is:

$$B = \begin{pmatrix} (2-6) & (6-0) & (3-0) \\ (4-4) & (10-0) & (10-0) \\ (12-4) & (30-12) & (5-6) \end{pmatrix}$$

$$\therefore B = \begin{pmatrix} -4 & 6 & 3 \\ 0 & 10 & 10 \\ 8 & 18 & -1 \end{pmatrix}$$

The transpose of $B(B^{\mathrm{T}})$ is:

$$B^{\mathrm{T}} = \begin{pmatrix} -4 & 0 & 8 \\ 6 & 10 & 18 \\ 3 & 10 & -1 \end{pmatrix}$$

The matrix $\begin{pmatrix} d & -b \\ -c & a \end{pmatrix}$ in equation (4.23) was the adjoint of $\begin{pmatrix} a & b \\ c & d \end{pmatrix}$.

We may now write the general expression for the inverse of matrix A as:

$$A^{-1} = \frac{\text{Adj } A}{|A|} \tag{4.33}$$

Problems 4.8

Obtain the determinants and thence the inverse form of the following matrices:

4.8.1

$$A = \begin{pmatrix} 3 & 4 & 1 \\ 2 & 5 & 2 \\ 4 & 1 & 3 \end{pmatrix}$$

4.8.2

$$A = \begin{pmatrix} 1 & 2 & 3 \\ 1 & 3 & 3 \\ 1 & 2 & 4 \end{pmatrix}$$

4.8.3

$$A = \begin{pmatrix} 1 & 2 & 3 \\ 1 & 3 & 5 \\ 1 & 5 & 12 \end{pmatrix}$$

4.9 Relation to coordinate geometry and solution of simultaneous equations

We can now begin to draw a parallel between the coordinate systems in two or three dimensions (x,y) or (x,y,z) and simple matrices, where matrix elements correspond to coordinate intersections. The coordinates (1,2) in a two-axis graph correspond directly to the element i_{12} in a matrix, except that columns in the graph are in an upward sense rather than the downward sense of matrices. Similarly we noted earlier that equations involving x and y or x, y and z can be represented in matrix form. The parallel should be a very clear one by this stage. For example, we can rewrite the equation, $y = 5x+8$ as:

$$(5 \quad -1)\begin{pmatrix} x \\ y \end{pmatrix} = (-8)$$

in matrix form. Now if we consider two simultaneous equations each including terms in x and y,

$$5x+5y = 15 \tag{4.34}$$

and

$$6x+2y = 14 \tag{4.35}$$

we can rewrite these as:

$$\begin{pmatrix} 5 & 5 \\ 6 & 2 \end{pmatrix}\begin{pmatrix} x \\ y \end{pmatrix} = \begin{pmatrix} 15 \\ 14 \end{pmatrix} \tag{4.36}$$

in matrix form. If we let

$$A = \begin{pmatrix} 5 & 5 \\ 6 & 2 \end{pmatrix} \quad \text{and} \quad k = \begin{pmatrix} 15 \\ 14 \end{pmatrix} \quad \text{and} \quad X = \begin{pmatrix} x \\ y \end{pmatrix}$$

then we can rewrite the matrices as $AX = k$. Multiplying each side by the inverse of A, A^{-1}, we now have:

$$A^{-1}AX = A^{-1}k$$

so that

$$IX = A^{-1}k, \quad \text{since } AA^{-1} = I$$

and thus

$$X = A^{-1}k \tag{4.37}$$

The solution of the equations can thus be obtained by multiplying k by the inverse of A. Now the inverse of A from equation (4.24) is:

$$A^{-1} = \frac{1}{10-30}\begin{pmatrix} 2 & -5 \\ -6 & 5 \end{pmatrix}$$

so that:

$$x = -\frac{1}{20}\begin{pmatrix} 2 & -5 \\ -6 & 5 \end{pmatrix}k$$

$$= \begin{pmatrix} -0{\cdot}1 & 0{\cdot}25 \\ 0{\cdot}3 & -0{\cdot}25 \end{pmatrix}\begin{pmatrix} 15 \\ 14 \end{pmatrix} = \begin{pmatrix} 2 \\ 1 \end{pmatrix}$$

or $x = 2$ and $y = 1$.

Perhaps the reader's initial reaction to all this is that the solution of simultaneous equations is much simpler by the more conventional methods outlined in chapter 3! Apart from demonstrating that the solution of simultaneous equations is possible by matrices, the matrix solution of such equations is useful when attempting this task using computer methods, where variable arrays may be used to store the simultaneous equations in matrix form, as shown above. For the simultaneous solution of a large number of equations each containing a large number of unknowns, generally very tedious by more conventional arithmetic methods, the above matrix technique for solution can easily be programmed for computer usage. Herein lies its main usefulness. However, because the technique is useful in this context and because we must take the view that nothing should be used from within the mathematical or statistical sphere without some knowledge of the individual operations concerned we are going to solve the example below where we have three simultaneous equations in three variables. Let us define the equations initially as:

$$3x+2y+z = 14 \tag{4.38}$$

$$2x+y-4z = 4 \tag{4.39}$$

$$x-y-z = 0 \tag{4.40}$$

We must now find the inverse of the matrix, using techniques from sections 4.7 and 4.8.

$$A = \begin{pmatrix} 3 & 2 & 1 \\ 2 & 1 & -4 \\ 1 & -1 & -1 \end{pmatrix} \tag{4.41}$$

We have in fact already found the determinant of this matrix (from equation (4.32)), and:

$$A = -22$$

The matrix of cofactors of (4.41) is, remembering to apply the alternate sign rule:

$$\begin{pmatrix} +(-1-4) & -(-2-(-4)) & +(-2-1) \\ -(-2-(-1)) & +(-3-1) & -(-3-2) \\ +(-8-1) & -(-12-2) & +(+3-4) \end{pmatrix}$$

which becomes:

$$B = \begin{pmatrix} -5 & -2 & -3 \\ +1 & -4 & +5 \\ -9 & +14 & -1 \end{pmatrix}$$

$$B^{\mathrm{T}} = \begin{pmatrix} -5 & +1 & -9 \\ -2 & -4 & +14 \\ -3 & +5 & -1 \end{pmatrix}$$

and therefore the solution for X is given by:

$$x = -\frac{1}{22} \begin{pmatrix} -5 & +1 & -9 \\ -2 & -4 & +14 \\ -3 & +5 & -1 \end{pmatrix} \begin{pmatrix} 14 \\ 4 \\ 0 \end{pmatrix} = \begin{pmatrix} 3 \\ 2 \\ 1 \end{pmatrix}$$

or $x = 3$, $y = 2$ and $z = 1$. The reader may check the solutions using the technique for the solution of simultaneous equations in chapter 3.

Problems 4.9

Solve the following sets of simultaneous equations using matrix algebra and check your answers by solving the equations simultaneously in the normal way.

4.9.1
$$3x-4y+2z = 19$$
$$2x+y-4z = 2$$
$$x+5y+z = 7$$

4.9.2
$$2y-3x+4z = 10$$
$$x = y+z-4$$
$$2z+y-6x = 23$$

4.9.3
$$3x+4z-2y = 6$$
$$2y-3x-4z = 6$$
$$3z+2y-10x = 1$$

4.10 Eigenvectors and eigenvalues

Perhaps one of the most important uses of matrices lies in their use in certain methods of statistical analysis. The technique whereby we take the 'principal components' to explain the variance of data in a multivariate situation is the most well known. Principal components analysis attempts to explain the overall variance with respect to new axes, such that the first principal component accounts for as much variance as possible, and each successive component after this accounts for progressively less variance. The total variance is thus entirely accounted for by the sum of each of the variances from each component. The technique and also therefore the nature of eigenvectors and eigenvalues which have to be calculated, is probably best explained with reference to graphs of a simple bivariate relationship shown in figure 4.1.

The graphs shown in figure 4.1(a) to (c) represent a simple two-way relationship between two variables; their names and what they represent are of no importance to this discussion. In the first example, in figure 4.1(a), we have a situation where the correlation between the two variables is zero, and the equiprobability contours (lines of equal probability) assume a series of concentric circles about the mean point of each variable represented on the graph. If we assume that both sets of data are normally distributed (see chapter 7) about means of zero in each case, and that the axes in figure 4.1 represent multiples of the standard deviations, then we can see in figure 4.1(a) that for a value of x equal to the mean less one standard deviation, there is a whole range of corresponding values of y, whose probabilities of being associated with the x value are related to the probabilities on the normal curve (see chapter 7). The second part of the figure (figure 4.1(b)) shows a situation where the two variables are related with a correlation coefficient of about 0·5. As we have seen already in chapter 3, there will be two least-squares regression lines associated with such a correlation. The angle between these regression lines has the cosine whose value equals that of the correlation coefficient. Since there is a better correlation between the variables we see that the lines of equiprobability are

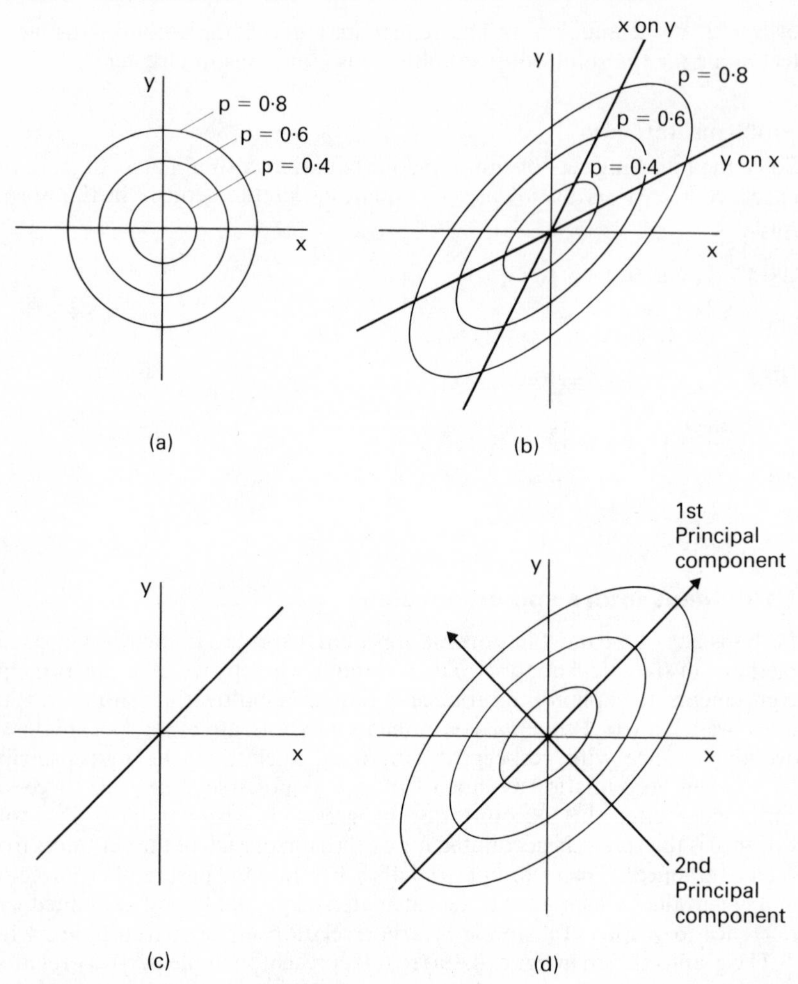

Figure 4.1 Regression lines, equiprobability contours and principal components: (a) $r = 0{\cdot}0$; (b) $r = 0{\cdot}5$; (c) $r = 1{\cdot}0$; (d) principal components.

accordingly distorted into an ellipse whose long axis lies between the two regression lines, along the reduced major axis, and whose short axis is at right-angles to it. These two axes are the first and second principal components (figure 4.1(d)). For information we see that the situation where we have perfect correlation in figure 4.1(c), the regression lines and therefore the principal components are coincident. Clearly the number of principal components is the same as the number of variables whose covariance is being examined. We shall restrict our attention to the problem of the determination of the principal components where we have only two axes: where two variables only are to be considered.

Each principal component has an equation which is defined in terms of the eigenvector (its direction) and its eigenvalue (normally represented by λ) which reflects the variance explained by each component. Here we are resolving vectors into two directions at right-angles. The eigenvalues constitute a diagonal matrix:

$$\begin{pmatrix} \lambda_1 & 0 \\ 0 & \lambda_2 \end{pmatrix}$$

The problem is to rotate the graph axes to transform the original data matrix into another which uses the principal components as axes. We saw in chapter 3 that rotation of axes through an angle θ gave new coordinates as follows:

$$x = x_1 \cos\theta - y_1 \sin\theta \qquad \text{(from (3.21))}$$

and

$$y = x_1 \sin\theta + y_1 \cos\theta \qquad \text{(from (3.22))}$$

Thus we can write in matrix form:

$$\begin{pmatrix} x \\ y \end{pmatrix} = \begin{pmatrix} \cos\theta & -\sin\theta \\ \sin\theta & \cos\theta \end{pmatrix} \begin{pmatrix} x_1 \\ y_1 \end{pmatrix} \tag{4.41}$$

For example to rotate axes through 42 degrees where sin 42 = 0·6691 and cos 42 = 0·7432, the point $P_1(1,2)$ becomes P (x,y) after rotation:

$$\begin{pmatrix} x \\ y \end{pmatrix} = \begin{pmatrix} 0{\cdot}7432 & -0{\cdot}6991 \\ 0{\cdot}6691 & 0{\cdot}7432 \end{pmatrix} \begin{pmatrix} 1 \\ 2 \end{pmatrix}$$

$$= \begin{pmatrix} -0{\cdot}5950 \\ 2{\cdot}1560 \end{pmatrix}$$

We can therefore write that after rotation $P_1 \quad (1,2) \rightarrow P \quad (-0{\cdot}595,\ 2{\cdot}156)$.

For any rotation certain factors remain 'invariant' regardless of the magnitude of the rotation. For example, if before rotation we have four points on a graph each depicting a corner of a geometric figure, then after rotation although the positions of each of the points has moved the geometric shape is retained. The shape is thus invariant. The eigenvector is another, special case, of an invariant. If we take matrix A and multiply it by the vector X, such that the result is itself a scalar multiple of the vector, or:

$$AX = \lambda X \tag{4.43}$$

then λ is the eigenvalue, and the vector X, the eigenvector. Expanding on equation (4.43) we have:

$$\begin{pmatrix} a & b \\ c & d \end{pmatrix} \begin{pmatrix} x \\ y \end{pmatrix} = \lambda \begin{pmatrix} x \\ y \end{pmatrix} \tag{4.44}$$

or

$$\begin{aligned} (a-\lambda)x + by &= 0 \\ cx + (d-\lambda)y &= 0 \end{aligned} \tag{4.45}$$

or

$$\begin{pmatrix} a-\lambda & b \\ c & d-\lambda \end{pmatrix} X = 0 \tag{4.46}$$

Now the equations in (4.45) are homogeneous (their right-hand sides are zero). For any pair of simultaneous equations which are homogeneous the determinants have the value zero, unless they are to have the solutions $x = y = 0$. Thus it follows that:

$$\begin{vmatrix} a-\lambda & b \\ c & d-\lambda \end{vmatrix} = 0 \tag{4.47}$$

or
$$(a-\lambda)(d-\lambda)-bc = 0$$
This equation is known as the *characteristic equation* of the matrix, from which the eigenvalues are obtained, and these results are substituted into equations (4.45) to give the eigenvector.

Let us now find the eigenvalues and eigenvectors of a simple 2×2 matrix, where:

$$A = \begin{pmatrix} \frac{1}{3} & 2 \\ \frac{1}{3} & 0 \end{pmatrix}$$

We have from equation (4.47):

$$(\tfrac{1}{3}-\lambda)(0-\lambda)-\tfrac{2}{3} = 0$$
$$\therefore \lambda^2-\tfrac{1}{3}\lambda-\tfrac{2}{3} = 0$$
or
$$3\lambda^2-\lambda-2 = 0$$

so that using the expression for finding the solution to quadratic equations (section 5.6, equation (5.32)) we have:

$$\lambda = \frac{1 \pm \sqrt{(1+24)}}{6}$$
$$= 1 \text{ or } -\tfrac{2}{3}$$

Substituting back in equations (4.45) we have:

for $\lambda = 1$
$$(\tfrac{1}{3}-1)x+2y = 0$$
and
$$\tfrac{1}{3}x+(0-1)y = 0$$

or in both cases; $y = \frac{1}{3}x$, the first eigenvector;

for $\lambda = -\frac{2}{3}$
$$(\tfrac{1}{3}+\tfrac{2}{3})x+2y = 0$$
and
$$\tfrac{1}{3}x+(0+\tfrac{2}{3})y = 0$$

or in both cases; $y = -\frac{1}{2}x$, the second eigenvector.

This example has been chosen as a deliberately elementary case to illustrate the application of equations (4.43) to (4.47) in the derivation of the eigenvectors and eigenvalues of a simple 2×2 matrix. We are approaching this problem as a mathematical one rather than a statistical one and can afford to take such a simple example. The aim has been to illustrate what eigenvectors and eigenvalues are, not to compute an actual example as it might occur as a result of statistical analysis of field data. If we were conducting a principal components analysis on field data in a multivariate situation the matrix for which we should be obtaining eigenvalues and eigenvectors would be the matrix of correlation coefficients. There would be as many coefficients as there were data pairs and the correlation matrix would reflect this. In addition the eigenvalues in the case of principal components analysis are the 'principal component scores' in the analysis and their sum will equal the number of variables used. There will be as many such scores as there are variables in the principal component analysis. Thus if we are conducting the analysis on data for which there are four variables, then we should have four eigenvalues, one for each component in each dimension, and their combined value would also equal four. Suppose we had computed scores of:

$$\lambda_1 = 2{\cdot}56$$
$$\lambda_2 = 0{\cdot}65$$
$$\lambda_3 = 0{\cdot}42$$
$$\lambda_4 = 0{\cdot}37$$

then the first principal component is the most important, and it accounts for 64·0% of the variance ($2{\cdot}56 \times 100/4$). The second accounts for 16·25%, the third, 10·5% and the fourth, 9·25% of the total variance. The sum of all four is the total number of variables involved.

As a simplification for use in the determination of eigenvalues and vectors, we can show that since:

$$\begin{pmatrix} a-\lambda & b \\ c & d-\lambda \end{pmatrix} = \begin{pmatrix} a & b \\ c & d \end{pmatrix} - \lambda \begin{pmatrix} 1 & 0 \\ 0 & 1 \end{pmatrix}$$

then:

$$|A - \lambda I| = 0 \tag{4.48}$$

another representation of the characteristic equation.

Any further development of the application of matrix algebra, eigenvalues and eigenvectors and the like to statistical techniques, is beyond the scope of this book, and the enthusiastic student or one forced by circumstances, is referred to the large number of good advanced statistical texts on the market, dealing specifically with principal components analysis. In particular the work by Kendall (1965) is almost the definitive text, but Hope (1968) also covers much the same ground. If the direct application to geography is required then the recent short text by Daultry (1976) is to be recommeded.

Problems 4.10

Determine the eigenvalues and eigenvectors for the following matrices:

4.10.1

$$A = \begin{pmatrix} 2 & 4 \\ 6 & 8 \end{pmatrix}$$

4.10.2

$$A = \begin{pmatrix} -4 & -2 \\ -1 & -3 \end{pmatrix}$$

4.10.3

$$A = \begin{pmatrix} 3 & 2 \\ 1 & 0 \end{pmatrix}$$

4.11 Matrices and networks

The major use of matrices within physical geography is as an aid to computation and in their use in principal components analysis. An example of their application to probability studies can be found in chapter 7, and the only significant remaining area within the subject where matrices may be used is in the field of networks. Within physical geography networks do not possess the complexity of for example, transport nets encountered in the human branch of the subject. Perhaps the easiest way to develop the topic is through stream networks. The early work on stream networks conducted by Horton (1945) and Strahler (1957) ordered streams according to the number and the magnitude of absorbed tributaries. We may represent stream patterns in matrix form according to the number of *links* (i.e. tributaries) converging upon each point of confluence (*nodal points*). Simple matrices may be constructed to demonstrate the network either in terms of links or in terms of nodes. A simple idealized stream network is shown in figure 4.2. Links are represented by the

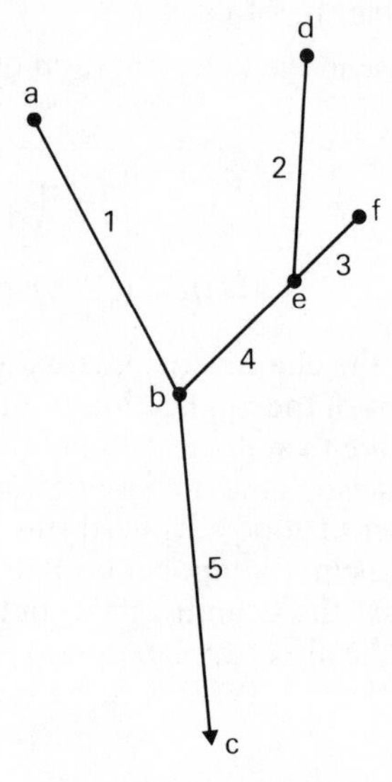

Figure 4.2 Idealized stream network showing nodes and links.

numbers one to five and nodes are indicated by lower case letters, a to f. The link and node matrices which represent this stream network are shown below.

$$\begin{array}{c} \\ 1 \\ 2 \\ 3 \\ 4 \\ 5 \end{array} \begin{array}{c} \begin{array}{ccccc} 1 & 2 & 3 & 4 & 5 \end{array} \\ \begin{pmatrix} 0 & 0 & 0 & 1 & 1 \\ 0 & 0 & 1 & 1 & 0 \\ 0 & 1 & 0 & 1 & 0 \\ 0 & 1 & 1 & 0 & 1 \\ 1 & 0 & 0 & 1 & 0 \end{pmatrix} \end{array} \qquad \begin{array}{c} \\ a \\ b \\ c \\ d \\ e \\ f \end{array} \begin{array}{c} \begin{array}{cccccc} a & b & c & d & e & f \end{array} \\ \begin{pmatrix} 0 & 1 & 0 & 0 & 0 & 0 \\ 1 & 0 & 1 & 0 & 1 & 0 \\ 0 & 1 & 0 & 0 & 0 & 0 \\ 0 & 0 & 0 & 0 & 1 & 0 \\ 0 & 1 & 0 & 1 & 0 & 1 \\ 0 & 0 & 0 & 0 & 1 & 0 \end{pmatrix} \end{array}$$

In the link matrix a zero indicates that there is no direct link between the respective tributaries, and one indicates that a direct link exists, so we can see that 2 is directly linked to 3 and 4 but not to 1 and 5. For the node matrix a similar notation is used, so that for example node d is directly linked to node e, but to none of the others, and node b is linked to a, c and e. Both matrices are symmetric, but clearly this symmetry could never be the case for a stream network, bearing in mind the simple expedient of water being unable to run uphill! Thus we must somehow indicate that linkages in one direction are not possible. The convention used to illustrate this, and to make the matrices non-symmetric, is that rows represent 'from' a, b, c and so on, and the columns indicate 'to' a, b, c and so on. Thus since we have linkages possible only from 1 to 5, or from f to e, we lose some of the linkages in each matrix. The two matrices are thus:

$$\begin{array}{c} \\ 1 \\ 2 \\ 3 \\ 4 \\ 5 \end{array} \begin{array}{c} \begin{array}{ccccc} 1 & 2 & 3 & 4 & 5 \end{array} \\ \begin{pmatrix} 0 & 0 & 0 & 0 & 1 \\ 0 & 0 & 0 & 1 & 0 \\ 0 & 0 & 0 & 1 & 0 \\ 0 & 0 & 0 & 0 & 1 \\ 0 & 0 & 0 & 0 & 0 \end{pmatrix} \end{array} \qquad \begin{array}{c} \\ a \\ b \\ c \\ d \\ e \\ f \end{array} \begin{array}{c} \begin{array}{cccccc} a & b & c & d & e & f \end{array} \\ \begin{pmatrix} 0 & 1 & 0 & 0 & 0 & 0 \\ 0 & 0 & 1 & 0 & 0 & 0 \\ 0 & 0 & 0 & 0 & 0 & 0 \\ 0 & 0 & 0 & 0 & 1 & 0 \\ 0 & 1 & 0 & 0 & 0 & 0 \\ 0 & 0 & 0 & 0 & 1 & 0 \end{pmatrix} \end{array}$$

The sum of each column thus indicates the total number of tributaries draining into each receiving river: two each into links 4 and 5 and two each into nodes b and e. Changes in the network can be easily represented in the matrices by matrix addition and subtraction. The process can be extended to portray other river network characteristics, such as discharge, channel size and so on.

5
Non-linear functions

If we were to invent the geographically perfect world it is probable that we would opt for all relationships within it to be linear! This would make for considerable ease in the analysis of field observations. We restricted ourselves to such relationships in chapter 3. To be realistic however, whilst the understanding of very simple linear relationships assists us in our understanding of non-linear functions, the majority of relationships found within physical geography are not as straightforward. Most frequently the physical geographer is concerned with the fitting of functions to a scatter of points on a graph, and in many cases the resulting lines are far from linear. It is therefore of great importance for the aspiring physical geographer to have a good working knowledge of the main types of *standard* (non-linear) functions and their graphical expressions. Once the aspiring physical geographer has acquired these skills and is able to establish mathematical functions which at least approximate the trend of points across the graph, then the resulting equations will permit him to make an objective comparison between similar relationships from different data populations.

The ultimate aim is thus the fitting of functions to observed trends of points on a graph. However, before we can look in more detail at the process of curve fitting (section 5.6) we must first establish the identities of the various major types of standard function, their forms on a graph and the general forms of their equations. First of all though we shall need to define some terms and concepts which will help in the understanding of the equations themselves and the forms they produce. This involves a brief introduction to *differential calculus*, a branch of mathematics of which an understanding is crucial to the remaining chapters in this book. The reader is urged to study the chapter well, for of all the nine in the book this is almost certainly the most immediately relevant to a majority of problems in physical geography itself.

5.1 Essentials of differential calculus

Curvilinear relationships pose two major problems in particular when compared with elementary linear functions. First of all, their slopes are constantly changing, and secondly, since many curves reverse direction at least once the functions can yield more than one value of y for any single value of x, or vice versa. The most elementary curvilinear function, $y = x^2$, has two solutions of y for every value of x, since y may take the value $+\sqrt{x}$ or $-\sqrt{x}$.

The slope of a line such as $y = x^2$ (figure 5.2) increases as the value of x increases. The curve is concave upwards. In a curve such as $y = 2x^2$ the slope is both a function of the value of x and the constant, 2. Similarly for a curve such as $y = \sin x$ between $x = 0$ and $x = \pi$ (figure 2.3), the slope is again a function of x but this time it *decreases* as x increases. The curve is convex upwards. The problem thus arises of how we may derive the value of the slope of a curve at a given point. With the straight line it was simple because the slope was unchanging. In addition, because of the curvature in non-linear examples we must find some way of considering a sufficiently small segment of the curve such that the slope we are looking at may be considered constant at that point. A curve, mathematically speaking, can be thought of as being composed of a number of infinitesimally small straight-line sections placed end to end, each section having a different slope. Any curve can be approximated on a graph by drawing in straight lines between points on the curve (*chords*). The closer these points are together, the shorter are the chords and the greater is the precision with which the curve is approximated. Thus to determine the true slope of a curve at a point, the straight sections must be so small that we are considering the slopes of *tangents* to the curve, and not the slopes of chords to the curve. The tangent to a curve is a straight line which contacts the curve at only one point and does not cross it. In figure 5.1 AB represents a very small segment of a curve whose equation is, for the moment, immaterial. Line CD is a chord

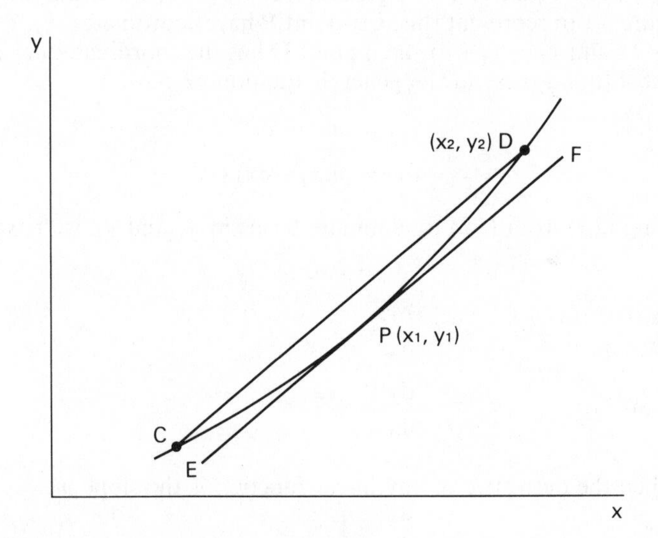

Figure 5.1 Determination of gradient of a curve.

joining two points on the curve, and line EF is the tangent to the curve at point P. Points P and D have coordinates (x_1, y_1) and (x_2, y_2) respectively. If we now imagine a whole series of very short chords such as CD extending end to end we can see how a curve may be approximated by them. It is by considering every curve as being made up of such infinitesimally small straight-line sections that we can approach the problem of determining the slope of a curve at a point.

Given the coordinates of D and P the slope of the line joining P to D is given by $(y_2 - y_1)/(x_2 - x_1)$ (from equation (3.1)). If we assume that the change in magnitude of each value of x or y along the line is very small between these points we can indicate these small changes by:

δy for a very small change in y (*not* δ times y)
δx for a very small change in x (*not* δ times x)

Thus the slope or gradient may be given by:

$$\frac{\delta y}{\delta x}$$

Strictly speaking of course, this value is in fact the slope of the chord PD, but as D moves closer to P then this slope value approaches the slope of the tangent EF. To represent this we write:

$$\frac{\mathrm{d}y}{\mathrm{d}x} = \lim_{x\to 0}\frac{\delta y}{\delta x}$$

This limit of $\delta y/\delta x$ as δx tends to zero is the *derivative.* When we calculate the derivative of a function we are approximating the slope of the straight line drawn at a tangent to the curve, and the process is known as *differentiation.* The slope of a straight line can now be demonstrated mathematically. Consider a linear equation in the general form, $y = mx+k$, using the section PD in figure 5.1 to represent this. Let point P have coordinates (x_1, y_1), since $x_2 = x_1+\delta x$ and $y_2 = y_1+\delta y$, then point D has the coordinates $(x_1+\delta x, y_1 + \delta y)$. Substituting these in the general equation we have:

$$y_1 = mx_1+k \tag{5.1}$$

$$(y_1+\delta y) = m(x_1+\delta x)+k \tag{5.2}$$

Subtracting (5.1) from (5.2) to eliminate terms in x_1 and y_1, we have:

$$\delta y = m\delta x$$

and thus

$$\frac{\delta y}{\delta x} = m$$

or as $\delta x \to 0$,

$$\frac{\mathrm{d}y}{\mathrm{d}x} = m \tag{5.3}$$

Clearly then the derivative of any linear function is the slope m.

For $y = x$, $\qquad \dfrac{\mathrm{d}y}{\mathrm{d}x} = 1$

and for $y = 6x$, $\qquad \dfrac{\mathrm{d}y}{\mathrm{d}x} = 6$ and so on.

Now let us assume that the function of the curve AB in figure 5.1 has the equation $y = ax^2+bx+c$. We shall consider other curves of this type at a later stage. Then, again using points P (x_1, y_1) and D $(x_1+\delta x, y_1+\delta y)$, we have:

$$y_1 = ax_1^2+bx_1+c \tag{5.4}$$

and

$$\begin{aligned}(y_1+\delta y) &= a(x_1+\delta x)(x_1+\delta x)+b(x_1+\delta x)+c \\ &= ax_1^2+a(\delta x)^2+2ax_1\delta x+bx_1+b\delta x+c \end{aligned} \tag{5.5}$$

Subtracting (5.4) from (5.5):

$$y_1 - y_1 + \delta y = ax_1^2 + a(\delta x)^2 + 2ax_1\delta x + bx_1 + b\delta x + c - ax_1^2 - bx_1 - c$$
$$\therefore \ \delta y = a(\delta x)^2 + 2ax_1\delta x + b\delta x$$
$$= (2ax_1 + b)\delta x + a(\delta x)^2$$

or
$$\frac{\delta y}{\delta x} = 2ax_1 + b + a\delta x \tag{5.6}$$

However, as $\delta x \to 0$ the term $a\delta x$ also tends to zero and thus considering a general point (x, y) the derivative of $y = ax^2 + bx + c$ is given by:

$$\frac{dy}{dx} = 2ax + b \tag{5.7}$$

or
$$\frac{d}{dx}(ax^2 + bx + c) = 2ax + b$$

We see here clearly that the derivative of a non-linear function is a function itself of the value of x, and of any constant term which may appear in the equation. Thus for example, in the equation $y = 6x^2 + 3x + 5$, the derivative is given by:

$$\frac{dy}{dx} = 12x + 3 \tag{5.8}$$

The total derivative of an equation composed of a number of parts of different powers of x is thus the sum of the derivatives of the individual terms. For example, in equation (5.8) we see that, as for the linear function $y = 3x$ the derivative of $3x$ is 3, and that the derivative of $6x^2$ is $12x$ (that is, $2 \times 6x$), and that therefore:

$$\frac{dy}{dx} = \frac{d}{dx}(ax^2) + \frac{d}{dx}(bx) + \frac{d}{dx}(c) \tag{5.9}$$

Certain common features emerge regarding the determination of the derivatives of powers of x. In the above example $y = ax^2$ had the derivative $dy/dx = 2ax$. If the function had been $y = ax^3$, then the derivative would have been $3ax^2$. Note that in each case the constant in front of the x term is multiplied by the original power of x, whilst the power itself is reduced by one. In fact the same occurs for $y = ax$ and $y = a$, except that in these two cases their derivatives are a and zero respectively, since in the first case the original power of x was one and in the second it was zero ($a = ax^0$). A general form for the derivative of a function $y = ax^n$ thus emerges as:

$$\frac{dy}{dx} = anx^{n-1} \tag{5.10}$$

This most useful general derivative should be memorized. The derivatives of other functions of x, involving logarithms or trigonometric ratios are given at a later stage in this chapter when the form of such functions is considered.

As an application of simple differentiation we may consider what happens to a body when it is dropped from rest. Neglecting air resistance, the body will accelerate towards the ground at $9{\cdot}81\ \mathrm{m\,s^{-2}}$ (the acceleration due to gravity denoted by g), indicating that after the first second of fall it will have a velocity of 9·81 m s^{-1}, after the second, 19·62 m s^{-1} (2 × 9·81) and so on. Thus, during the first second it will have a mean velocity of 9·81/2 = 4·91 m s^{-1}; during the second, a mean velocity of (9·81 + 19·62)/2 = 14·72 m s^{-1} and during the third, a mean velocity of 24·53 m s^{-1}. After one second it will thus have covered 4·91 m, after the second, 4·91 + 14·72 = 19·62 m and after the third, 19·62 + 24·53 = 44·15 m. The relationship between distance fallen (s) and the time elapsed (t) may thus be represented by a curvilinear function of the form:

$$s = \tfrac{1}{2}gt^2 \tag{5.11}$$

where s is a power function of t. Using the rule developed in equation (5.10) we have:

$$\frac{\mathrm{d}s}{\mathrm{d}t} = 2.\tfrac{1}{2}gt^{2-1}$$

or

$$\frac{\mathrm{d}s}{\mathrm{d}t} = gt \tag{5.12}$$

In other words the *velocity* ($\mathrm{d}s/\mathrm{d}t$) of the falling object after t seconds is given by gt, acceleration multiplied by time. Thus if we now let $v = \mathrm{d}s/\mathrm{d}t$ and differentiate v with respect to t we obtain:

$$\frac{\mathrm{d}v}{\mathrm{d}t} = 1.gt^{1-1}$$

or

$$\frac{\mathrm{d}v}{\mathrm{d}t} = g \tag{5.13}$$

Hence we have arrived at the original acceleration. This indicates the constant acceleration g towards the earth's surface. This is borne out when we consider the dimensions in turn of distance, velocity and acceleration. For any equation representing a physical relationship the dimensions of the units on either side of the equation must balance. Virtually all units may be expressed in terms of three basic measures, mass (M), length (L) and time (T), regardless of the measuring system used (Imperial, SI, etc.). For example in the equation $s = \frac{1}{2}gt^2$, the dimensions of the left-hand side are L, and those of the right-hand side are $\mathrm{LT}^{-2}\mathrm{T}^2 = \mathrm{L}$. Further expansion of this topic may be found in Davidson (1978). Equally we can see that acceleration ($\mathrm{d}v/\mathrm{d}t$) is a *rate of change* of velocity with time. Thus differentiation gives us a means of determining rates of change of the dependent variable with respect to the independent variable in an expression. The example of lapse rates given in chapter 3 was an expression of a rate of change of temperature with altitude ($\mathrm{d}T/\mathrm{d}z$). Acceleration is the *second derivative* of distance and the *first derivative* of velocity, the relationships indicating such derivatives may be written:

$$a = \frac{\mathrm{d}^2s}{\mathrm{d}t^2} = g$$

$$v = \frac{\mathrm{d}s}{\mathrm{d}t} = gt$$

$$s = \tfrac{1}{2}gt^2$$

The reverse process to differentiation, *integration*, is considered separately in chapter 8. The particular equations of motion above have been developed by Statham (1976) to produce a model for rockfall.

An alternative notation for derivatives which is commonly used in many mathematical texts, and which, for convenience, we shall use in chapter 9 is as follows. If we let $y = f(x)$ without specifying further the relationship, then:

$$\frac{dy}{dx} = f'(x) = y'$$

$$\frac{d^2y}{dx^2} = f''(x) = y''$$

and $$\frac{d^3y}{dx^3} = f^{(3)}(x) = y^{(3)} \text{ and so on}$$

Thus if $f(x) = 3x^5 + 2x$, $f'(x) = 15x^4 + 2$, $f''(x) = 60x^3$ and $f^{(3)}(x) = 120x^2$ and so on.

Problems 5.1

Obtain the first and second derivatives (dy/dx and d^2y/dx^2) of the following functions:

5.1.1 $y = 5x^3$
5.1.2 $y = 4 - 2x^2$
5.1.3 $3y = x^3 - 2x + 5x^2$
5.1.4 $y = 12x + 3$
5.1.5 $y = x(x+1)(x-1)$
5.1.6 $y = 1/x^2$
5.1.7 $y = 2x^{-4}$
5.1.8 $y = \dfrac{x(x^2 - x - 2)}{x+1}$
5.1.9 $y = \frac{1}{3}x^3 - x^2 + 4$
5.1.10 $y = \frac{1}{3}x^3 - 3x^2 + 9x + 4$

5.2 Characteristics of the standard functions

5.2.1 Power functions

We are now in a position to describe the main types of standard function in detail. The fitting of curves to observed relationships can be a difficult and lengthy task, since resultant functions are often highly complex. An outline of the basic approach to curve fitting follows in section 5.6. However, a number of computer programs also exist which have been specifically designed for the purpose. Many major computer manufacturers offer 'subroutine packages' in their software which permit the fitting of basic curves; for example in the ICL 1900 series there is a scientific subroutine entitled F4CFORPL. A number of basic curve types can be identified though, according to certain mathematical and morphological criteria. If such identification is possible then the task of curve fitting without recourse to programs or lengthy calculation is made easier. It is therefore useful at this stage simply to list the functions and their associated graphical forms. Specific examples from the field of physical geography are found at the end of this chapter in section 5.7.

Power functions take the general form $y = x^n$. At their most simple, they are linear relationships: when $n = 0$ (i.e. the function $y = 1$) or when $n = 1$ (the function $y = x$). However, for values of n other than zero or one curves are produced. Three different types of curve result depending upon whether $n > 1{\cdot}0$: or $-0{\cdot}\dot{9} \leq n \leq 0{\cdot}\dot{9}$: or $n < -1{\cdot}0$. We shall consider each of these in turn.

The first category where $n > 1{\cdot}0$ is illustrated in figure 5.2 with respect to the integer powers of x (2, 3 etc.). Since in the majority of derived relationships within physical geography we are concerned only with both positive x- and y-values, it is perhaps most useful to restrict our attention to the forms of curves within the first quadrant, at least for the present. All curves in this category share two points (0,0) and (1,1), since one raised to any power is always one, and similarly for zero. As n increases above 1·0 however, the curvature of the line between the origin and (1,1) increases also. For very high values of n therefore, x must approach a relatively high decimal fraction before the value of y becomes appreciable. A whole spectrum of curves sharing the points (0,0) and (1,1) can thus be envisaged as n increases, with the most acute form apparent for very high values of n, and the most open form associated with low values of n, culminating in the straight line $y = x^1$.

If we now turn our attention to the remaining three quadrants of the graph in figure 5.2 we can clearly see that for *even* values of n the form of the curve in the first quadrant is *mirrored* in the second quadrant. However, for *odd* values of n (e.g. $y = x^3$) the curve is *inverted and mirrored* such that it occupies the

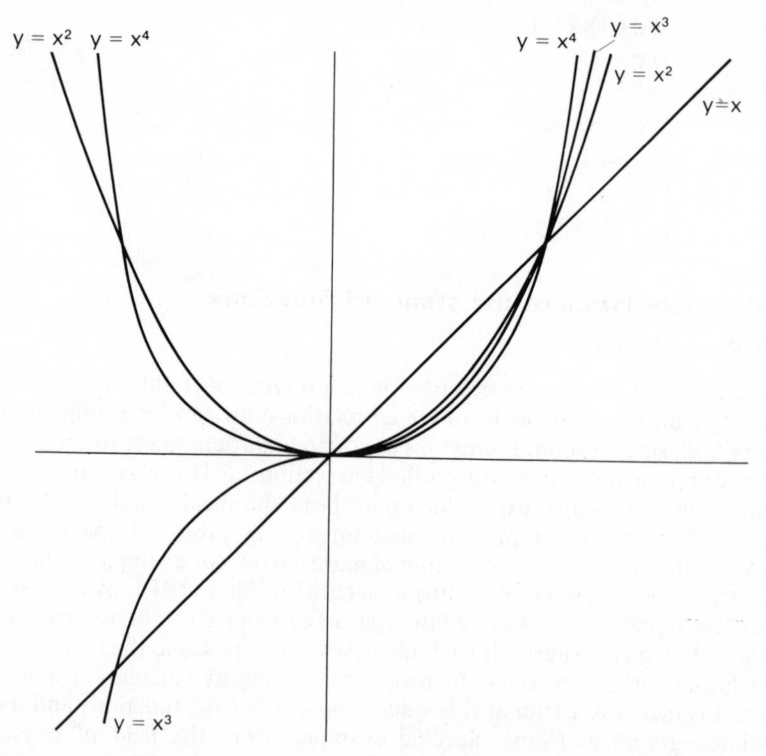

Figure 5.2 Curves of the form $y = x^n$, $n > 1{\cdot}0$.

third quadrant. This feature is of course simply explained in terms of the sign rules for powers explained in chapter 1. Negative values raised to odd powers retain their negative values, whilst negative values raised to even powers become positive. The curve for $y = x^2$ is a special case in that it describes a *parabola*. Further details of the parabola and other specific curve forms such as the circle and ellipse, may be found for example, in Flanders and Price (1973). Certain mathematical definitions exist for a curve of this form, and in mathematics textbooks these are generally given for a parabola about the x-rather than the y-axis. Such parabolas fall within the next category of power function where n lies between zero and $0{\cdot}\dot{9}$.

The expression of, for example, $y = x^{\frac{1}{2}}$ is simply an alternative way of writing the function $y^2 = x$, and thus the form taken by curves whose n value lies between the limits of $-1{\cdot}0$ and $+1{\cdot}0$ is that of symmetry about the x-axis rather than only the y-axis or both, which is the case for the first and last category of power functions. In fact the equation $y^2 = 4ax$ is the general expression used for the parabola. A parabola may be defined as the *locus* of a point which moves so that it maintains a constant ratio between its distance from a fixed point (F, the *focus* in figure 5.3), and a fixed straight line (DD', the *directrix*). Thus in figure 5.3 PM/PF is a constant. The form of a parabola may be determined by varying the ratio PM/PF. This relationship determines that, for example in a car headlamp, the rays of light shining from F onto a parabolic surface will always be reflected parallel to the axis bisecting the parabola; the principle of parabolic reflectors.

The curves for the functions of $y = x^{\frac{1}{2}}$ and $y = x^{\frac{1}{4}}$ are shown in figure 5.4. Smaller values of n create a tighter fit of the curve about the x-axis. Again though, our main concern as physical geographers will be with the first

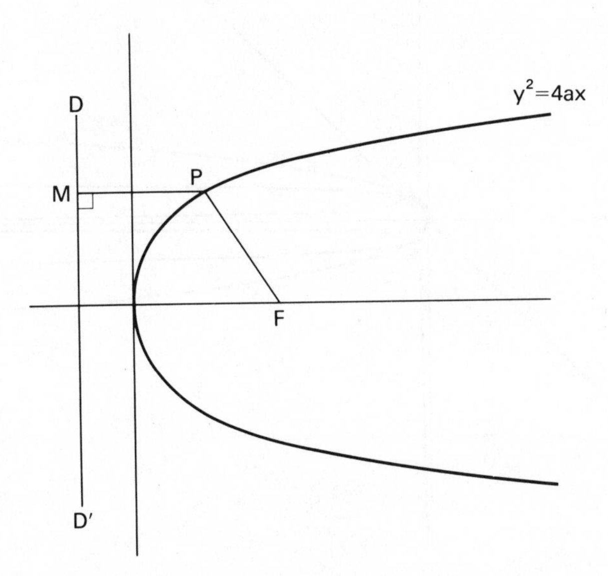

Figure 5.3 Definition of a parabola about the x-axis: $y^2 = 4ax$.

quadrant, and we can see here that there is a *decrease* of gradient in the first quadrant as x increases, rather than the reverse which was the case when $n > 1$. The extreme cases for positive values of n are enclosed within the 'curve' for $y = x$ and the line defined by $y = 1$, where $n = 0$. There is thus a continuation of the trend observed for functions when $n > 1$. For negative n (but $n > -1$) however, an important change in curve form takes place, although symmetry about the x-axis is retained. The graph of $y = x^{-1/2}$ is shown in figure 5.4. In this case, within the first quadrant, we see that the form of the curve is such that with increasing x the line approaches the x-axis, and with increasing y it approaches the y-axis. In fact a detailed consideration of the function clearly indicates that the curve never actually intersects the axes, although the curve and axes converge as infinity is approached. Curves which approach the axes but intersect them only at infinity are termed *asymptotic*. In these examples the asymptotes are actually the x- and y-axes, but there are many examples where asymptotes do not coincide with the axes and are not parallel to them. The geometric form defined by functions such as these is the *hyperbola*. All functions of the form $y = x^n$ where $n < 0$ (including those where n is less than -1) are *rectangular* hyperbolas, since their asymptotes are at right-angles, namely the x- and y-axes. Examples from within physical geography are given in sections 5.6 and 5.7.

Whilst the power functions in the second category introduced above were symmetrical only about the x-axis, once n becomes less than or equal to $-1{\cdot}0$ a different symmetry develops. The curves are still hyperbolic in form (see figure 5.5) in that their limbs approach the axes asymptotically, but parts of them

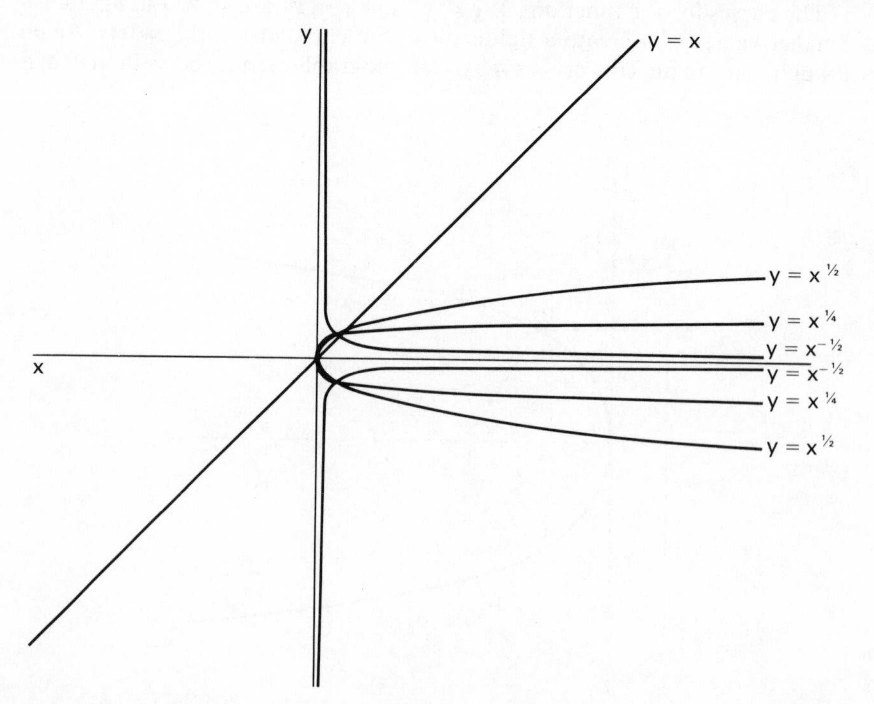

Figure 5.4 Curves of the form $y = x^n$, $-0{\cdot}9 \leq n \leq 0{\cdot}9$.

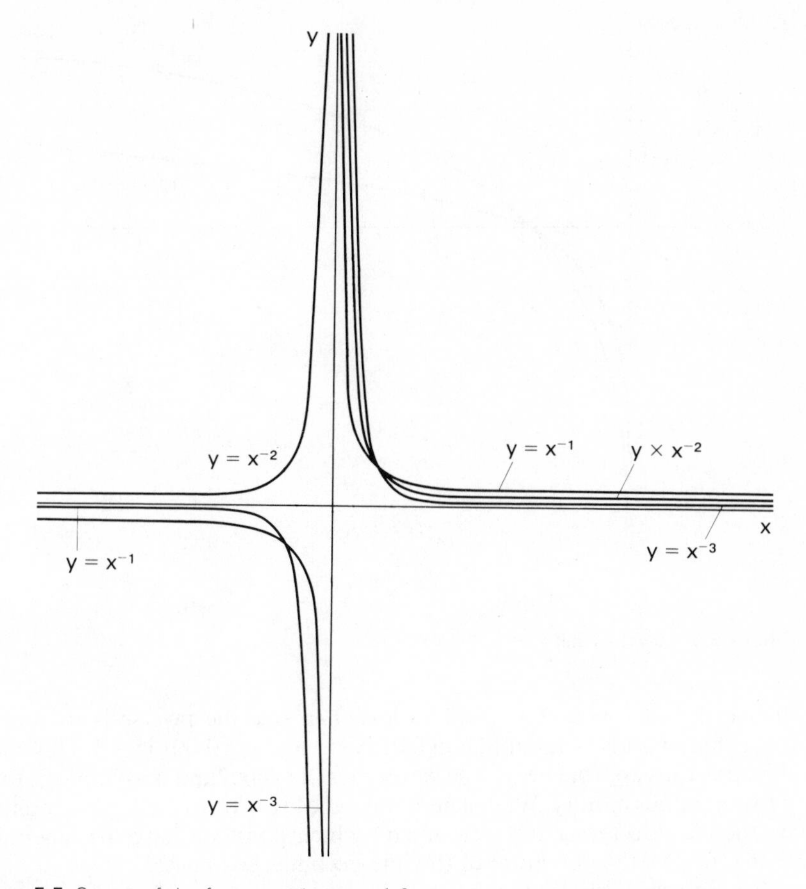

Figure 5.5 Curves of the form $y = x^n$, $n \leq -1{\cdot}0$.

occur in the second or third quadrants, rather than in the fourth quadrant. These are the inverse functions $y = 1/x$, $y = 1/x^2$, $y = 1/x^3$ and so on. Once x is less than zero (at which point y is infinity) odd powered functions yield negative values in both x and y, whilst even powered functions are characterized by negative terms only in x, as was the case when $n > 1{\cdot}0$.

5.2.2 Logarithmic and exponential functions

Logarithmic and exponential functions produce curves of a form which differ markedly from those produced by power and subsequent groups of functions. The form of the curve produced by the general function $y = \log_a x$ varies slightly as the base a varies. The form is perhaps best illustrated first of all using logarithms to the base ten, and the curve of $y = \log_{10} x$ is shown in figure 5.6. The logarithm of one to any base is always zero, and thus all curves of the form $y = \log_a x$ intersect at the common point (1,0). Using logarithms to the base ten the logarithm of 10 is 1·0, of 100 is 2·0 and so on. Thus as the value of x increases the value for $y = \log_{10} x$ increases, but at an increasingly slow rate: whilst the curve reaches $y = 1$ at $x = 10$, x must equal 100 before the curve

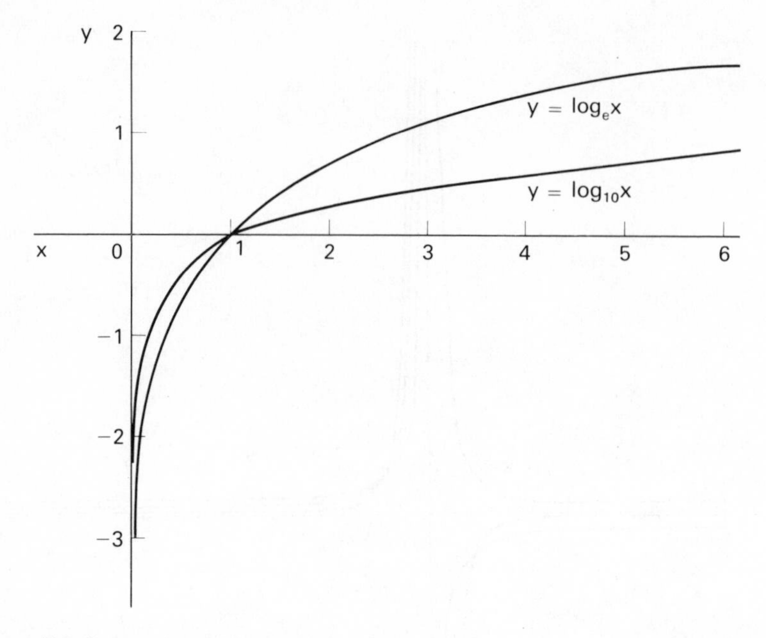

Figure 5.6 Curves of the form $y = \log_a x$.

intersects with $y = 2$. For x-values less than one the reverse is true as the logarithm of 0·1 is -1, but that of 0·01 is -2 and log 0·001 is -3. This end of the curve thus approaches but never reaches the y-axis and is asymptotic to the y-axis at minus infinity. We can never take the logarithm of a negative number. As the base decreases towards one (at which point we have the function $y = \log_1 x = 1$), the curvature of the line becomes less acute.

If we are concerned with the function $y = \log_e x$ we are effectively taking the function $x = e^y$, and thus the exponential function of the form $y = e^x$ is the same as $x = \log_e y$. Hence the form of the curve representing this function is very similar to the logarithmic function, except that it is asymptotic to the x-axis at minus infinity, and y is never less than zero. The curve of the function $y = e^{-x}$ is a mirror image about the y-axis of $y = e^x$ (figure 5.7). The form of the curve is distinctly flattened when the power x is multiplied by small decimal fractions, as for example, $y = e^{0 \cdot 1x}$.

The nature of the two types of curve means that the rule derived earlier for the calculation of the derivative from power functions cannot be applied. We can obtain the derivatives of the two functions in a similar way, but the methods used are beyond the mathematical scope required of this book. Thus, it is sufficient to give the general form of the derivative for each:

$$\frac{d(\log_a x)}{dx} = \frac{1}{x} \tag{5.14}$$

$$\frac{d(e^x)}{dx} = e^x \tag{5.15}$$

Very definite properties are clear in each case. In the first it is important to note

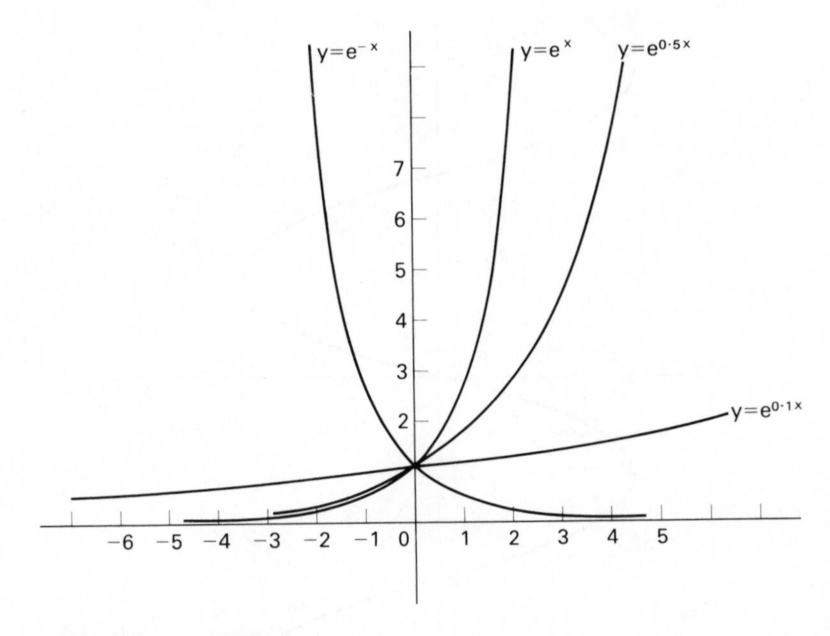

Figure 5.7 Curves of the form $y = e^{ax}$.

that the derivative is the same for any base of logarithm, and in the second that any derivative (first, second, third etc.) of the exponential function is always e^x.

5.2.3 Trigonometric and inverse trigonometric functions

The basic trigonometric functions have been introduced in chapter 2. These functions of sin x, cos x and their reciprocals take the form of a continuous and repeating waveform, with a wavelength of $x = 2\pi$ radians. The case of the tangent and the cotangent may not at first sight appear truly wave-like as they are discontinuous. This too was the case for the curves of some of the power functions. They do however, repeat at a regular interval of 2π radians. Examples of the graphs of the functions $y = \sin x$, $y = \cos x$ and $y = \tan x$ appeared in figures 2.3 and 2.4. As such the functions require little further description. Their inverse form is, however, useful from time to time. Instead of taking the function $y = f(x)$ we can write its inverse as $x = g(y)$. Thus in the context of the trigonometric functions we require to write $\sin y = x$ in a more conventional form. This becomes,

$$y = \sin^{-1} x \tag{5.16}$$

and is the *inverse sine function* (figure 5.8). It must be noted that we are not here using the suffix -1 to indicate that $y = 1/\sin x$. Clearly this would be the same as writing $y = \sec x$. The suffix is used in trigonometric functions purely to indicate (as in the above example) the *angle of y in radians whose sine is x*. Another term used in the same context, but restricting values of x to between $-\pi/2$ and $+\pi/2$, is that of the *arcsine*. Similarly we can have the arctangent. Neither of these functions will repeat, and they apply only between the two

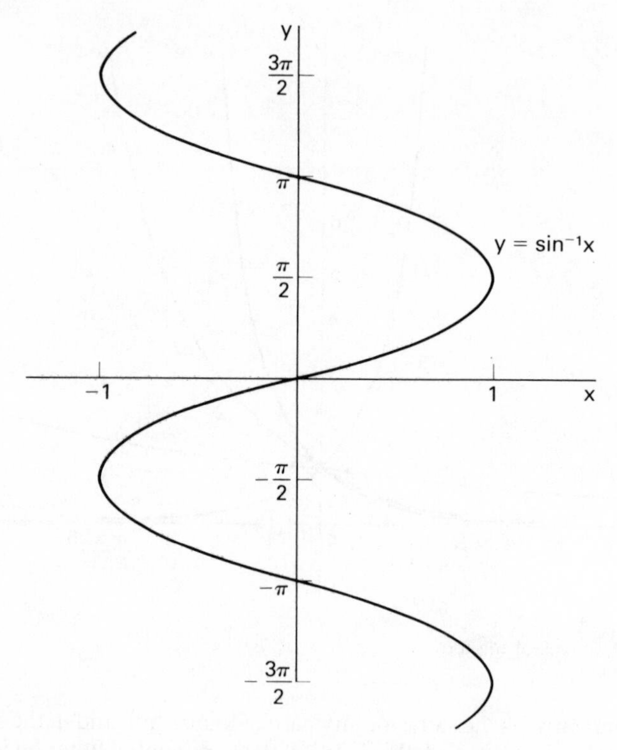

Figure 5.8 Curve of $y = \sin^{-1}x$, an inverse trigonometric function.

limits. However, true inverse trigonometric functions repeat and oscillate in the same way as trigonometric functions, but about the y-axis rather than the x-axis. The derivatives of each of the main inverse trigonometric functions are:

$$\frac{\mathrm{d}}{\mathrm{d}x}\sin^{-1}\left(\frac{x}{a}\right) = \frac{a}{\sqrt{(a^2 - x^2)}} \tag{5.17}$$

in only the first and fourth quadrants; otherwise multiply by -1.

$$\frac{\mathrm{d}}{\mathrm{d}x}\cos^{-1}\left(\frac{x}{a}\right) = \frac{-a}{\sqrt{(a^2 - x^2)}} \quad \text{(1st and 2nd quadrants)} \tag{5.18}$$

$$\frac{\mathrm{d}}{\mathrm{d}x}\tan^{-1}\left(\frac{x}{a}\right) = \frac{a}{a^2 + x^2} \tag{5.19}$$

Derivatives of the standard trigonometric functions are:

$$\frac{\mathrm{d}(\sin x)}{\mathrm{d}x} = \cos x \tag{5.20}$$

$$\frac{\mathrm{d}(\cos x)}{\mathrm{d}x} = -\sin x \tag{5.21}$$

$$\frac{\mathrm{d}(\tan x)}{\mathrm{d}x} = \sec^2 x \tag{5.22}$$

There is another group of functions which at first sight are apparently trigonometric in form, although they are rarely encountered in physical geography. These are the *hyperbolic functions* of the form $y = \sinh x$, $y = \cosh x$ and $y = \tanh x$. The values of sinh x and so on are in fact derived from certain combinations of exponential functions, in that:

$$\sinh x = (e^x - e^{-x})/2 \tag{5.23}$$

$$\cosh x = (e^x + e^{-x})/2 \tag{5.24}$$

$$\tanh x = (e^x - e^{-x})/(e^x + e^{-x}) \tag{5.25}$$

We can also define the inverse hyperbolic functions, and the derivatives of these and of the hyperbolic functions are:

$$\frac{d(\cosh x)}{dx} = \sinh x \tag{5.26}$$

$$\frac{d(\sinh x)}{dx} = \cosh x \tag{5.27}$$

$$\frac{d(\tanh x)}{dx} = \operatorname{sech}^2 x \tag{5.28}$$

$$\frac{d}{dx}\sinh^{-1}\left(\frac{x}{a}\right) = \frac{1}{\sqrt{(x^2+a^2)}} \tag{5.29}$$

$$\frac{d}{dx}\cosh^{-1}\left(\frac{x}{a}\right) = \frac{1}{\sqrt{(x^2-a^2)}} \tag{5.30}$$

$$\frac{d}{dx}\tanh^{-1}\left(\frac{x}{a}\right) = \frac{1}{a^2-x^2} \tag{5.31}$$

For (5.30) $x/a>1$ and $\cosh^{-1}(x/a)>0$: for $\cosh^{-1}(x/a)<0$ multiply the derivative by -1. For (5.31) $x^2<a^2$.

5.3 Combinations of functions and their derivatives

The three basic types of non-linear function, and the linear function in chapter 3, provide the basic tools of the trade with which the physical geographer may attempt to construct mathematical models for observed relationships. Unfortunately however, it is not always possible to obtain a satisfactory fit using the standard function forms given in the previous three sections. There is not the space in this chapter to include all possible combinations of functions with examples of their graphical form, but the reader will have gathered that it is possible at least to approximate the form of a curvilinear relationship simply by reference to the forms of the functions in sections 5.2.1 to 5.2.3. For oscillating forms attempts at fitting trigonometric or inverse trigonometric functions may be made. Where there is clearly a trend of increasing gradient with increase in magnitude of x, simple power functions where $n>1{\cdot}0$ or exponential functions may suffice. Similarly where slope is seen to decrease with increase in x, logarithmic or power functions where $0{\cdot}0<n<1{\cdot}0$, may provide an approximation to an observed trend. However, it is worth considering some of the variations in curves which can be produced by combining functions of different types in the same equation. There are few hard and fast rules relating the combination of functions to the form of the curve produced. The simple addition of a constant value is sufficient to vary the position of a function with respect to the axes. A constant it will be remembered will not figure in the derivative and thus the form of the curve produced is unchanged. The curve $y = x^2 + 6$ will thus 'pivot' at (0,6) rather than at the origin which is the case for $y = x^2$. The establishment of such a *stationary point*, where curves change direction, is very important in curve plotting and fitting, and is dealt with in detail in section 5.4. Variation in curve form can be produced by the addition of further terms in x. In many cases power functions can be added to form a string of terms all involving terms in x, for example: $y = 4x^4 + 5x^3 + x^2 - 4x + 5$. Such a function is referred to as a *polynomial* and the most basic polynomials contain terms in x of less than the third power. This basic type is the *quadratic*. Polynomials having their highest power of x as x^3 and x^4 are, respectively, called *cubic* and *quartic*. Two simple examples of such polynomials (one quadratic and the other cubic) appear in figure 5.9. For the quadratic function ($y = x^2 - x - 1$) we note that this is again a curve with only one stationary point, but that this time it is not coincident with the origin, and that the curve is also not symmetrical about this point. The cubic polynomial ($y = x^3 - 2x^2$) in the figure shows little immediate similarity with $y = x^3$. Two stationary points appear, only one of which is coincident with the origin. The first of these points is called the *maximum* stationary point, since at its apex there is a maximum value in y, and the second is called the *minimum* stationary point, where there is a minimum y-value. Note however, in each case that maximum and minimum do not necessarily indicate the greatest or least values which y attains throughout the whole curve. Where both maximum and minimum points coincide we have what is termed a *point of*

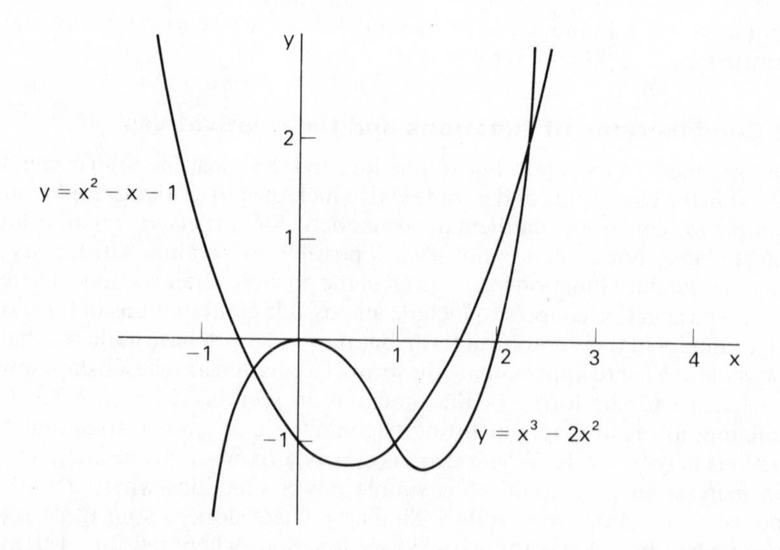

Figure 5.9 Curves of $y = x^2 - x - 1$ and $y = x^3 - 2x^2$.

inflexion. Two rules may now be expressed qualitatively. Firstly, for any polynomial, the curve will only pass through the origin if there is no constant term in its equation. Secondly, and most importantly, the number of stationary points on a curve is equal to the power of the highest-order term in the equation, *minus one*. So, for a quadratic polynomial we have only one such point, for a cubic polynomial, two stationary points, and for a quartic, three stationary points. Having said this however we must note that sometimes *all* stationary points coincide, where we have a point of inflexion.

Other combinations of functions are equally possible and it is impossible to show examples of every one. One of the most useful exercises the reader can carry out is to spend some time plotting the curves of different function combinations. Such an exercise will assist considerably in a deeper understanding of their workings. However, before this is possible it will help if we are able to establish exactly where stationary points occur.

5.4 The establishment of stationary points

From the brief treatment of polynomials in the previous section it is clear that stationary points of all three types are associated with infinitely short horizontal sections of the curves of the functions. Thus, for any function we may obtain firstly the x-coordinate of the point or points by equating the first derivative of the function to zero. The corresponding y-coordinate may then be obtained by substituting the x-value or values into the original equation. For example, if we take the function $y = x^2 - x - 1$ in figure 5.9 its first derivative is $2x - 1$. The x-coordinate of its stationary point is thus given by $2x = 1$, or $x = \frac{1}{2}$. Substituting $x = \frac{1}{2}$ into the original function we obtain $y = \frac{1}{4} - \frac{1}{2} - 1 = -1\frac{1}{4}$, and thus the coordinates of this minimum point are $(\frac{1}{2}, -1\frac{1}{4})$.

For higher-order polynomials, and particularly those involving powers higher than x^3, the solution for x in the equation produced by the first derivative can be an onerous task. The task involves the finding of factors of the equation, and very often these are difficult to obtain. For the solution of quadratic equations (produced when equating the first derivatives of cubic functions to zero), a formula exists into which we may substitute values from the equation in order to obtain the roots (solutions) of the equation. There will be two roots for every quadratic equation to be solved, three for every cubic and so on. At their most elementary quadratic equations may be solved by inspection for roots. For example, the equation $x^2 + 2x + 1 = 0$ may be factorized into $(x+1)(x+1) = (x+1)^2 = 0$, and there is only one solution for x, $x = -1$: both roots coincide. However, if we take the equation $x^2 + 4x - 12 = 0$, we may factorize this by inspection: $(x+6)(x-2) = 0$. The technique is basically one of experience and practice, plus a little trial and error! The roots of the equation $(x+6)(x-2) = 0$ are $x = -6$ and $x = 2$.

Now let us take the equation $2x^2 - 5x + 2 = 0$. This equation will not conveniently factorize into integer roots. This, and all similar quadratic equations may be solved by using the following formula:

$$x = \frac{b \pm \sqrt{b^2 - 4ac}}{2a} \tag{5.32}$$

where a, b and c are the constants in the general quadratic equation: $ax^2 + bx$

$+c = 0$. The solution for x in $2x^2 - 5x + 2 = 0$, is thus given as:

$$x = \frac{10 \pm \sqrt{25-16}}{4} = \frac{10 \pm 3}{4}$$

and thus $x = 3\frac{1}{4}$ or $x = 1\frac{3}{4}$. For higher-order equations the solution of the roots of x is rendered even more difficult. Unless the equation can be easily factorized by inspection, which is rare, the best action to take is to attempt an *approximate* solution by graphical means. This may be done by plotting the curve of the polynomial, and reading values of x off the graph at $y = 0$. For example, suppose we wish to find the stationary points of the function $y = x^4 - 3x^2 - x + 2$, then the first derivative of this is $4x^3 - 6x - 1$, and we are thus faced with the problem of solving the cubic equation $4x^3 - 6x - 1 = 0$. By plotting the function of $y = 4x^3 - 6x - 1$ and reading off values of x at $y = 0$ we can find the roots of the equation. First however we must obtain the stationary points of this new function. These are given by $12x^2 - 6 = 0$. The points thus occur at $x = \pm 1/\sqrt{2}$, i.e. $x = \pm 0{\cdot}71$. By substitution into the original cubic function we obtain turning points at $(0{\cdot}71, -3{\cdot}83)$ and $(-0{\cdot}71, 1{\cdot}83)$. The rest of the curve may now be plotted by obtaining values of y for x in the range of values determined by the points at which the curve intersects with the x-axis ($y = 0$). For the function the curve is drawn in figure 5.10 for $-2{\cdot}0 \leq x \leq +2{\cdot}0$, obtained from values calculated for table 5.1. We can now read off values of x at $y = 0$, which are $x = -1{\cdot}1$, $x = -0{\cdot}2$ and $x = 1{\cdot}25$, the approximate roots of

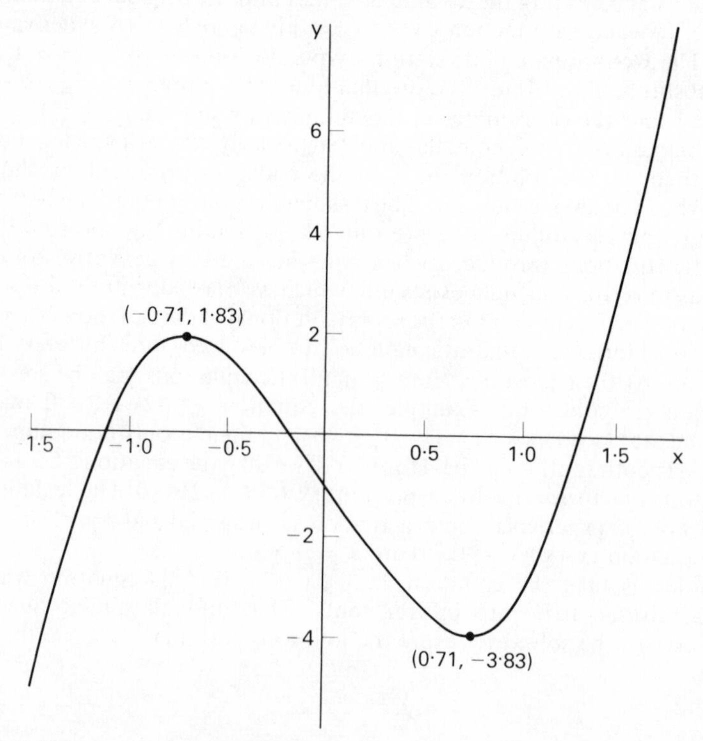

Figure 5.10 Determination of roots graphically, $y = 4x^3 - 6x - 1$.

Table 5.1 Points on the curve: $y = 4x^3 - 6x - 1$.

x	−2·0	−1·5	−1·0	−0·5	0·0	0·5	1·0	1·5	2·0
y	−21	−5·5	+1·0	+1·5	−1·0	−3·5	−3·0	+3·5	+19

$4x^3 - 6x - 1 = 0$. These can be checked by substitution. Clearly they are only approximations to the actual roots, but their accuracy may be increased by a more accurate plot of the curve of the function $y = 4x^3 - 6x - 1$.

It is clear from the graph plotted in figure 5.10 which stationary point is the maximum and which the minimum. However, the plotting of graphs of functions is a rather laborious means of determining which is which. We may use the second derivative of the function to distinguish between them. The second derivative of a function is an expression of the *rate of change of slope* of the curve. Clearly therefore once we have obtained the coordinates of the stationary points from the first derivative and the original function, if we determine the character of the rate of change of slope at these points we may very simply ascertain whether they are maximum or minimum points, or points of inflexion. If the value of the second derivative at the stationary point is *positive* then the gradient is *increasing* at that point and it must therefore be a *minimum* point. If on the other hand the second derivative is *negative*, gradient is *decreasing* and the point is a maximum. Where the value of the second derivative is zero we have a point of inflexion. Thus, in the previous example, the function $y = 4x^3 - 6x - 1$ has a maximum at $(-0{\cdot}71, 1{\cdot}83)$ since $d^2y/dx^2 = 24x = -17{\cdot}04$, and a minimum at $(0{\cdot}71, -3{\cdot}83)$ as $d^2y/dx^2 = 24x = +17{\cdot}04$.

We may now turn back to the original problem: that of deriving the stationary points for the function $y = x^4 - 3x^2 - x + 2$. The x-coordinates of the stationary points were given by the solutions for x in the cubic function $y = 4x^3 - 6x - 1$. Substitution of these values into the original quartic function yields the coordinates for the stationary points as $(-1{\cdot}1, 0{\cdot}9)$ for which the second derivative is positive, and thus it is a minimum, $(-0{\cdot}2, 2{\cdot}1)$ where d^2y/dx^2 is negative, a maximum, and $(1{\cdot}25, -1{\cdot}5)$ a further minimum point.

Problems 5.4

Before proceeding with this chapter the reader is urged to attempt the following exercises. Obtain the roots of the following quadratic equations:

(a) by inspection and factorizing

5.4.1 $x^2 - 4 = 0$
5.4.2 $x^2 + 16x + 64 = 0$
5.4.3 $2x^2 + 4x - 6 = 0$
5.4.4 $3x^2 + 4x + 1 = 0$
5.4.5 $1 - x^2 = 0$

(b) by using equation (5.32)

5.4.6 $3x^2 + 2x - 1 = 0$
5.4.7 $x^2 - 4 = 0$
5.4.8 $5x^2 - 7x - 2 = 0$
5.4.9 $-3x^2 + 4x - 1 = 0$
5.4.10 $3x - 2x^2 = 4$

Solve for the roots of x in the following equations by graphical means:

5.4.11 $x^3 - x^2 - 9x + 9 = 0$

5.4.12 $10x^3 + 23x^2 + 5x - 2 = 0$

Obtain the stationary points for each of the functions given in problems 5.1, and state whether they are maxima, minima or points of inflexion.

5.5 Derivatives of products and quotients

The plotting of more complicated types of function combinations (such as $y = x \log x$) again involves the calculation of stationary points to permit a relatively accurate graph to be drawn. The derivatives of such functions are themselves combinations of the derivatives of the individual functions and of the functions themselves. In the previous examples of function combinations we have seen only the addition of function derivatives. This proved to be a relatively simple if sometimes lengthy task. Many equations contain combinations of individual standard functions within them such that we require the derivative of the *product* (e.g. $y = x \log x$) or the *quotient* (e.g. $x/\log x$). In the general case if u and v are both functions of x, such that $y = u(x)v(x)$, then the derivative of the product is given by:

$$\frac{dy}{dx} = u(x)\frac{dv}{dx} + v(x)\frac{du}{dx} \tag{5.33}$$

and for the quotient, $y = u(x)/v(x)$

$$\frac{dy}{dx} = \frac{v(x)\dfrac{du}{dx} - u(x)\dfrac{dv}{dx}}{v(x)v(x)} \tag{5.34}$$

So for the function $y = x \log x$ we have:

$$\frac{dy}{dx} = \frac{1}{x}x + \log x = 1 + \log x$$

Similarly to find the derivative of $y = 3x^2 \sin x$, we have by substituting $u(x) = 3x^2$ and $v(x) = \sin x$ in equation (5.33):

$$\frac{d}{dx}(3x^2 \sin x) = 3x^2 \cos x + 6x \sin x$$

whilst for $y = \log x \tan x$,

$$\frac{dy}{dx} = \log x \sec^2 x + (\tan x)/x$$

Now for the quotient example above, we have:

$$\frac{d}{dx}\left(\frac{x}{\log x}\right) = \frac{\log x - x\dfrac{1}{x}}{\log^2 x}$$

$$= \frac{1}{\log x}\left(1 - \frac{1}{\log x}\right)$$

Again, if $y = \sin x/\cos x$ ($y = \tan x$, from which we know the derivative is $\sec^2 x$, equation (5.24)), then from equation (5.35) we have:

$$\frac{d}{dx}\left(\frac{\sin x}{\cos x}\right) = \frac{\cos^2 x + \sin^2 x}{\cos^2 x} = \sec^2 x$$

A final rule which can be used in finding the derivatives of combined functions is that known as the *function of a function* or *differentiation by substitution rule.* Consider the case of the following function:

$$y = (\cos x + \sin x)^3$$

By letting $z = \cos x + \sin x$ such that:

$$y = z^3$$

then
$$\frac{dy}{dz} = 3z^2$$

and
$$\frac{dz}{dx} = -\sin x + \cos x$$

Now,
$$\frac{dy}{dx} = \frac{dy}{dz}\frac{dz}{dx}$$

and therefore,

$$\frac{d(\cos x + \sin x)^3}{dx} = 3(\cos x + \sin x)^2(\cos x - \sin x)$$

Similarly we may find the derivative of $y = \log(\sin x)$ by letting $z = \sin x$ so that $dy/dz = 1/z$ and $dz/dx = \cos x$, and thus:

$$\frac{d}{dx}(\log(\sin x)) = \frac{1}{z}\cos x = \cot x$$

Still more complicated functions can be differentiated by applying more than one of the above rules to the same expression, and in this way derivatives of most functions can be obtained. Take for example the function $y = \tan 3x^3 + 2x \cos x$. The derivative of the first term ($\tan 3x^3$) may be found by the function of a function rule, and the derivative of the second term by the differentiation of a product method (equation (5.34)). For the first term let $z = 3x^3$, so that $y = \tan z$, thus we have

$$\frac{dy}{dx} = \frac{dz}{dx}\frac{dy}{dz} = 9x^2 \sec^2 3x^3$$

and for the second term let $y = 2x \cos x$, and so

$$\frac{dy}{dx} = 2\cos x + 2x(-\sin x) = 2(\cos x - x \sin x)$$

Thus:

$$\frac{d}{dx}(\tan 3x^3 + 2x \cos x) = 9x^2 \sec^2 3x^3 + 2(\cos x - x \sin x)$$

It is now very important that the reader familiarizes himself with the various techniques of differentiation and with the standard differentials given in this chapter. The problems below provide a convenient means of meeting this need. Further practice may be gained by attempting problems set in any text on elementary calculus (e.g. Knight and Adams 1975 or Stephenson 1973).

Problems 5.5

Differentiate:

5.5.1 $y = 5(2x^2 + \sin x)$
5.5.2 $y = 2 \sin^{-1} 2x$
5.5.3 $y = 3 \log x^2$
5.5.4 $y = x^{-2} \log x$
5.5.5 $y = 2 \sin x \cos x$
5.5.6 $y = \sec^2 x$ (let $z = \sec x$ and $u = \cos x$)
5.5.7 $y = \tan(\log x)$
5.5.8 $y = e^{2x} \log(x^2 + 1)$
5.5.9 $y = x \log x / \sin x$
5.5.10 $xy = 5 \cos x + x^2 \log x$

5.6 Curve fitting

The task of obtaining curvilinear regression lines of best fit to observed distributions of data is only slightly more demanding than that of fitting best fit linear regression equations. Ultimately, the statistical methods used and their prerequisites remain the same as those given in chapter 3. Lines can be drawn in by eye, although the derivation of the equations for them is more difficult than it was for straight lines. The fitting of polynomials or functions involving long strings of terms is beyond the scope of this book. However, the enthusiastic and confident reader is referred to virtually any text on numerical analysis which contains a chapter on curve fitting (e.g. Hildebrand 1974 or Khabaza 1965). One aspect of curve fitting which is best left to later in the book however, is that of the fitting of curves to periodic functions, as in harmonic analysis and Fourier analysis (chapter 8).

Thus, restricting our attention to simple curve types, one of the most useful means of determining the equation of a simple curve to fit given points is to 'straighten' the curve by plotting for example, one or both variables with reference to a logarithmic axis. At the end of chapter 3 mention was made of the possibilities of 'converting' non-linear relationships into a linear form by substituting for example, $t = \log x$ in an equation of the form $y = a \log x + c$ are constants, and then plotting y against $\log t$. Many observed relationships in physical geography approximate this form of function and they may be plotted as straight lines simply by utilizing special 'semi-logarithmic' graph paper in which one axis, normally the x-axis, is logarithmic in form: that is the scale has been distorted such that in the most simple case the plot of $y = \log x$ appears as a straight line (figure 5.11(a) and (b)). This should not be thought of as cheating! It is a quite legitimate means of better handling a function, and if it is possible then a line may be fitted to a scatter of points by means of one of the regression techniques outlined in chapter 3. Probably the best known of relationships in physical geography are the series of graphs which can be plotted relating drainage characteristics (length, drainage area and so on) to stream order, initiated by Horton (1945) and later developed by Strahler

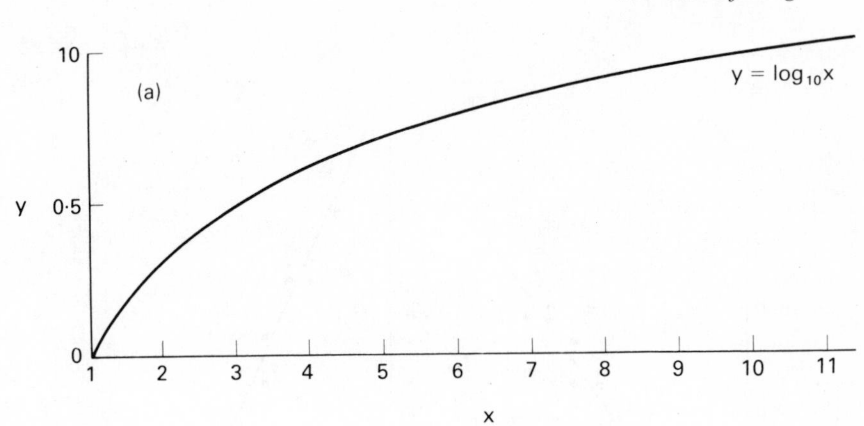

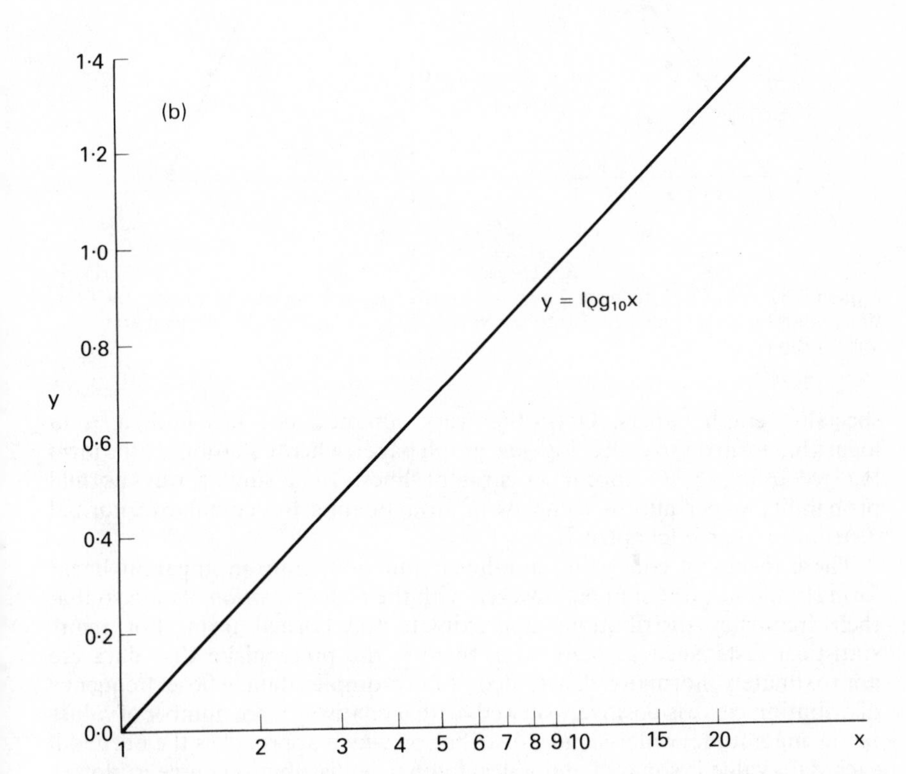

Figure 5.11 The graph of $y = \log_{10} x$: (a) on ordinary graph axes; and (b) on a semi-logarithmic graph.

(1957) and others. It must however be pointed out that these relationships may be observed for any natural branching system and cannot easily be associated with process in geomorphology, so that their usefulness is somewhat limited (King 1966). As examples however they are relatively convenient, well known and simple, and two such graphs are shown in figure 5.12. Further similar means of straightening plots of functions are available using a large number of

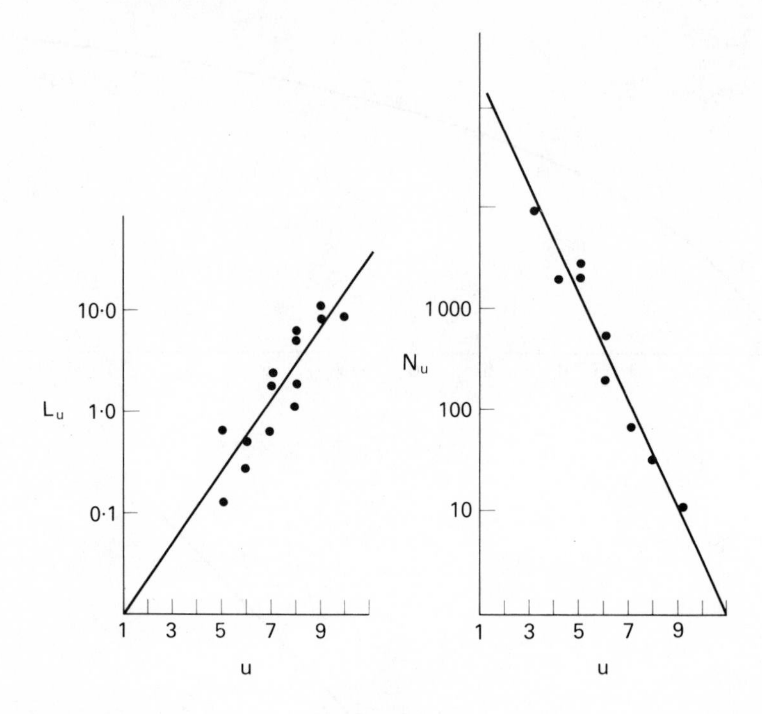

Figure 5.12 Graphs of the mean length of streams ($\bar{L}_u$) and number of streams (N_u) for stream order u for arroyos near Santa Fe, New Mexico. After Leopold, Wolman and Miller (1964).

specialist graph papers. One other very common one has both axes in logarithm form on so called 'log–log' graph paper, where equations of the form $\log y = a \log x + C$ appear as straight lines. In a similar way normal probability paper affords a means of straightening the cumulative normal distribution curve (chapter 7).

These means of converting non-linear functions into an apparent linear form should not be confused however, with the *transformation* of data so that their frequency distributions approximate the normal form. For many statistical tests, such as regression, there is the prerequisite that data are approximately normally distributed. For example, data whose frequency distribution curve is positively skewed (with a relatively large number of values of low magnitude) will assume a distribution which approaches the normal if each data value is squared and a new frequency distribution curve is plotted using these values. Similarly, for data which show a marked negative skewness, with a relatively large number of values of high magnitude, a normal distribution may be brought about by taking the square root or logarithm of each value. A brief definition of the normal distribution curve and examples of skewed distributions and transformations is given in appendix 5, section 4. The normal distribution curve itself is treated mathematically in chapter 7. Frequently of course, the statistical need for transformation of one or both variables in a relationship, may render a previously non-linear function linear.

A first stage in fitting a curve to data is often to identify any similarity between the curve plotted on ordinary graph paper and any of the examples of standard functions given earlier in the chapter. The process is one which consists very much of trial and error, unless one of the available curve-fitting programs is used. Consider for example, the relationship which has been observed between maximum rainfall intensity (I_t) and duration (t), for a large number of storms over periods of several years (McCallum 1959). A curve typical of such a relationship is shown plotted in figure 5.13. The visual agreement between points and the line is quite close, but the curve in this form is not very useful for predictive purposes, and could not easily be used to extrapolate maximum intensities over very long (greater than 12-hour) or very short (less than 5-minute) durations. The latter may well be very important in assessing the drainage capacity needed in the area to which the data apply. Over the data period the highest recorded intensity in five minutes for this site in a three year period (in west Wales) was 52·8 mm h^{-1}. Clearly the curve as drawn suggests that maximum intensities over even shorter periods, say of one minute, will be very much greater, but instrumental imprecision does not permit reliable estimates of intensities to be made over such short time periods. The curve is asymptotic to the y-axis (representing rainfall intensity). Similarly the other limb of the curve is apparently asymptotic to the x-axis (duration in hours). Comparison of the form of this curve with examples of the standard functions in figures 5.2 to 5.8 suggests that the function representing the relationship has characteristics in common with negative power functions, of the form $y = kx^{-n}$. Such functions can be conveniently rewritten in the form

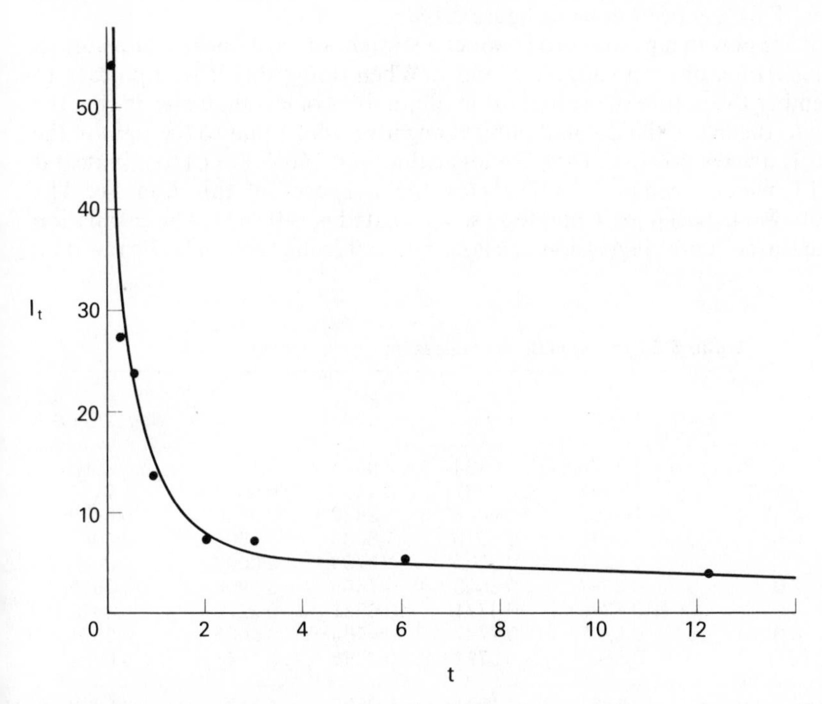

Figure 5.13 Graph of maximum rainfall intensity I_t (mm h^{-1}) over duration t (h) for Lampeter Wales, 1973–1975, plotted on ordinary graph paper.

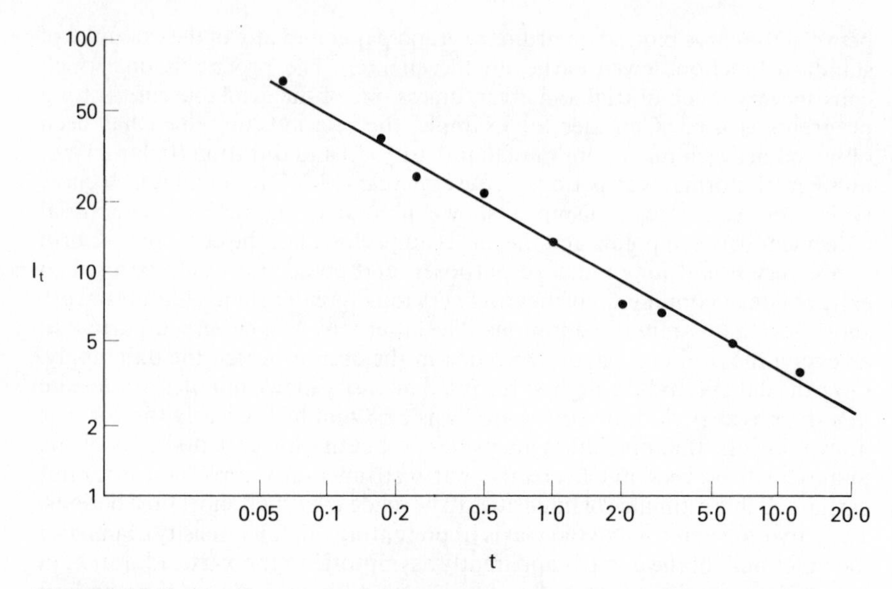

Figure 5.14 Graph of maximum rainfall intensity over duration plotted on double logarithmic graph paper. Details as for figure 5.13.

$\log y = \log k - n \log x$, which is similar to the general form for a linear function and will approximate a straight line if plotted with respect to double-logarithm paper. This has been done in figure 5.14.

We are now in a position to conduct a straightforward linear regression on the *logarithm* of each variable I_t and t. When doing this it is important to remember the nature of the logarithm of numbers of less than one, in that the value to the left of the decimal point is negative, whilst that to the right of the point is always positive. Thus the logarithm of 0·16667 hour (10 minutes) is $\bar{1}{\cdot}2219$, which becomes $-0{\cdot}7821$ for the purposes of this exercise. The correlation between $\log I_t$ and $\log t$ so calculated is $-0{\cdot}8837$. The calculation of the least-squares regression line is carried out using table 5.2. The raw data

Table 5.2 Semi-logarithmic regression for the function $I_t = kt^{-n}$.

I_t	t	y $\log I_t$	x $\log t$	y^2 $(\log I_t)^2$	x^2 $(\log t)^2$	xy $(\log I_t \log t)$
52·8	0·0833	1·7726	−1·0794	2·9674	1·1651	−1·8594
36·5	0·1667	1·5623	−0·7781	2·4408	0·6054	−1·2156
25·6	0·25	1·4082	−0·6021	1·9830	0·3625	−0·8479
22·8	0·5	1·3578	−0·3010	1·8436	0·0906	−0·4087
13·4	1·0	1·1271	0·0000	1·2704	0·0000	0·0000
7·4	2·0	0·8692	0·3010	0·7555	0·0906.	0·2616
6·8	3·0	0·8325	0·4771	0·6931	0·2276	0·3972
4·9	6·0	0·6902	0·7782	0·4764	0·6056	0·5371
3·6	12·0	0·5563	1·0792	0·3095	1·1647	0·6004
		10·1262 Σy	−0·1251 Σx	12·7397 Σy^2	4·3122 Σx^2	−2·5353 Σxy

values of rainfall intensity (I_t) and duration (t) are shown in the first two columns. The columns headed x and y contain the logarithms of I_t and t respectively, so that $y = \log I_t$ and $x = \log t$. The final three columns give values of x^2, y^2 and xy, with their sums at the foot of each column. The mean value of x is $-0{\cdot}0139$ and that of y is $1{\cdot}1251$, so that substituting in for Σxy, Σx, Σy, Σx^2 and $(\Sigma x)^2$ in equation (3.30) we have for an equation of the form $y = mx + k$:

$$m = \frac{9 \times (-2{\cdot}5353) - (-0{\cdot}1251)(10{\cdot}1262)}{9 \times 4{\cdot}3122 - (-0{\cdot}1251)^2}$$

$$= \frac{-21{\cdot}55}{38{\cdot}79} = -0{\cdot}56$$

Substituting for $\bar{x}$, $\bar{y}$ and m in equation (3.31) we also have:

$$k = 1{\cdot}1251 - (-0{\cdot}56)(-0{\cdot}0139)$$
$$= 1{\cdot}1173$$

Our least-squares regression equation is thus:

$$y = -0{\cdot}56x + 1{\cdot}1173$$

but $y = \log I_t$ and $x = \log t$ and therefore:

$$\log I_t = -0{\cdot}56 \log t + 1{\cdot}1173$$

Taking antilogarithms we have:

$$I_t = (\text{antilog}\, 1{\cdot}1173)t^{-0{\cdot}56}$$

and thus the equation of the curve in figures 5.13 and 5.14 is:

$$I_t = 13{\cdot}1t^{-0{\cdot}56} \qquad (5.36)$$

Up to this point we have conveniently used a linear regression technique which has assumed the data to be normally distributed without having established this fact. In this particular example, the duration data were strongly positively skewed in their untransformed state, but showed a better distribution when the logarithm of each value had been taken. Generally, curve fitting by this method is relatively haphazard, and the example chosen here illustrates one of the few examples whereby the relationship closely resembles a simple, standard function. There are many other examples which can only be approximated by such a simple function, and an ability to fit more complicated functions can only be obtained with experience and knowledge of what curve forms various function combinations produce, again emphsizing the importance of working through the problems set in this chapter.

5.7 Some further examples of non-linear relationships

Even the most cursory glance through the numerous journals and texts concerned with the various aspects of physical geography will reveal that many non-linear relationships, particularly within fluvial geomorphology and hydrology, follow the pattern set by the previous example, in that they are double-logarithmic in form. Most of these functions contain powers ranging between -1 and $+1$. Amongst them is a number in which y is inversely proportional to the *square root* of x. Equation (5.36) could be closely approximated for example, to $I_t = 13/\sqrt{t}$. A general function of the form:

$$i = a + bt^{-1/2} \tag{5.37}$$

has been found to relate the infiltration rate (i) into a soil after time (t) over a broad range of conditions (Philips 1957–1958). The application of the curve to a given set of conditions depends upon the minimum rate at which water will pass through the soil once it is saturated. This is dependent upon soil type and is represented by the constant a. The second term in the equation indicates that the variation in infiltration rate can be expressed solely in terms of the time elapsed after the start of infiltration, compounded by a constant factor, b, reflecting pre-existing soil moisture. If the soil was almost saturated at the start of infiltration then this term will be very small, as the infiltration rate will be approaching that for a saturated soil when $b = 0$. The curve produced by this equation takes on the form shown in figure 5.15 and is asymptotic to $i = a$. Similar relationships, although without the first constant term, often emerge when relating river runoff per unit area to contributory area. Thus, in the production of normal maximum flood curves for example, that for the 1952 Lynmouth flood (Dobbie and Wolf 1953), we find functions of the form, $R = kA^{-1/2}$. In this case Dobbie and Wolf found that:

$$R = 3350A^{-0 \cdot 5} \tag{5.38}$$

where R was the runoff per 1,000 acres in cusecs (cubic feet per second) and A was the contributory drainage area in thousands of acres.

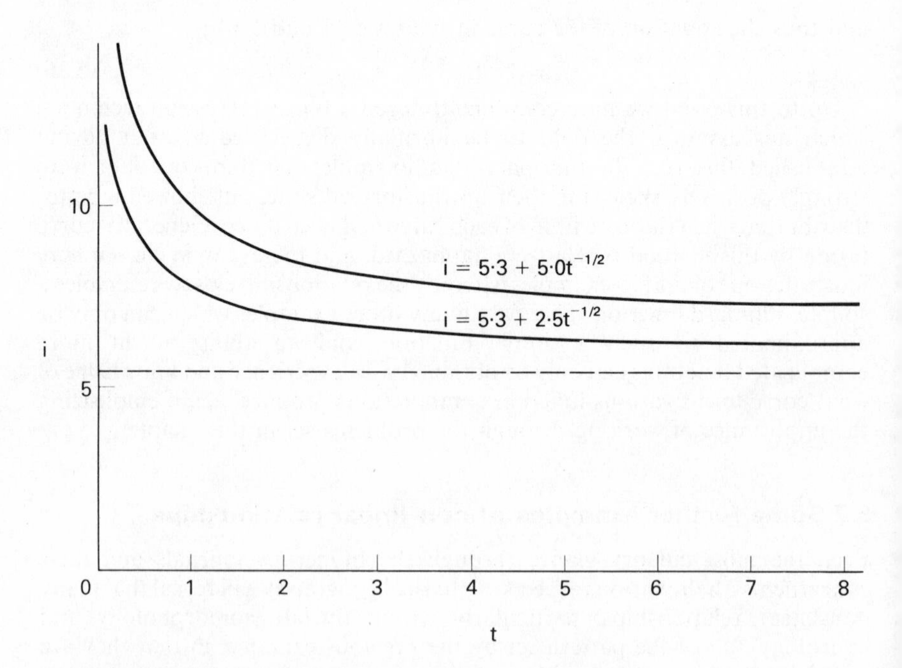

Figure 5.15 Graphs of the function $i = a + bt^{-\frac{1}{2}}$, for $a = 5 \cdot 3$ and $b = 2 \cdot 5$ and $b = 5 \cdot 0$, illustrating the effect of a high initial soil moisture content (lower b-value) and low initial soil moisture content (higher b-value) on the rate of infiltration.

Equations of a very similar form can also be found in hydrometeorological as opposed to hydrological studies. In most convectional storms for example, the general spatial distribution of total rainfall depth about a central maximum is such that totals decline radially from the centre of the storm. In many areas of the world the relationship between mean rainfall depth (P) for a storm of given duration and storm area approaches the form:

$$\bar{P} = a - bA^{1/2} \tag{5.39}$$

where the first constant term (a) is the central maximum precipitation for the storm, and b expresses the rate of rainfall decrease radially away from the storm centre. This particular relationship has been found to hold true for many parts of the United States (Huff and Neill 1957). Other studies concerned with deriving a similar relationship in the United States (for example, Miami Conservation District 1936) have found that over relatively large areas the function contains an extra, exponential term. These are of the general form:

$$\bar{P} = P_0 e^{-kA^n} \tag{5.40}$$

where P_0 is equivalent to the constant term a in equation (5.39), and k and n are constants associated with a given storm or series of storms of the same duration. One example of a derived equation of the above form for parts of the northern United States is:

$$P = 16e^{-0{\cdot}0883A^{0{\cdot}24}} \tag{5.41}$$

This previous example was typical of some of the more complex functions which are occasionally encountered. A stage in curve fitting can be reached however, whereby an increase in the complexity of the function describing a relationship yields but a small increase in the goodness of fit of the curve to the observed data points. For the purpose of comparison it may well be argued that a relatively simple function approximating the observed relationship is the most desirable: even in some cases a linear function. Many relationships in physical geography apparently result in simple bivariate functions generally adopting an approximate power function form. It must be realized however that in most cases this is an oversimplification of the case in two senses. Firstly, in the sense mentioned above, the derived function is only an approximate description of reality, and secondly, some of the apparent discrepancy between the observed relationship and the calculated function may be a result of interference by other independent factors. Some functions which have been derived within fluvial geomorphology for example, clearly also exclude a number of other important independent variables in what must inevitably be a multivariate situation. A third point to make is that where functions describing observed relationships give a linear graph on double-logarithm paper the apparent goodness of fit may deceive the reader into thinking that a very high degree of correlation exists between the variables concerned. Consideration of the axes of such graphs clearly indicates that the degree of modification afforded is such that relatively large deviations of points from the line for higher values can be effectively disguised. Care is sometimes needed therefore in the interpretation of such relationships. Two examples of such power functions where the agreement is relatively good however, are those derived for the relationship between stream length (L) and drainage area (A), such as for example:

$$L = 1{\cdot}4A^{0{\cdot}6} \tag{5.42}$$

developed by Hack (1957), and that between the greatest recorded rainfall depths over the globe as a whole (R) and duration (D), often cited in climatology texts (for example, Smith 1972, Linsley and Franzini 1972), of:

$$R = 16{\cdot}6D^{0{\cdot}47} \tag{5.43}$$

Finally a number of examples exist which may be portrayed as straight lines on semi-logarithmic paper: functions of the form $y = f(\log x)$. This is again particularly so in the field of fluvial geomorphology. One of the first forays into the world of mathematical modelling within this field was carried out by Green (1935) when he attempted to fit a general curve to the longitudinal profile of the River Mole in England. His intention was to attempt to relate to a general curve former base-levels as represented by terrace remnants down the valley, and in doing so reconstruct the possible former profile and infer the sea level to which it was 'graded'. The closeness of fit is shown in figure 5.16. The equation is of the general form:

$$y = a - k \log(p - x) \tag{5.44}$$

where y is the altitude in feet above sea level, x is the distance in miles from the mouth of the river, p is the total river length, and a and k are constants which in the case of the Mole were 241·5 and 65 respectively. The curve is in fact asymptotic towards the inferred former sea levels at one end and to the y-axis at the other—conditions which in practice could never be reached. Similar types of semi-logarithmic relationships were also found by Ruhe (1969) in the comparison of loess deposits down valleys in Iowa, USA, and by Horton (1945) in his pioneering work on the morphometry of stream networks. For example, reference to figure 5.12 reveals that the plot of mean stream length for a given order and the order of streams themselves is very nearly a perfect semi-logarithmic relationship:

$$\bar{L}_u = f(\log u) \tag{5.45}$$

where $\bar{L}_u$ is the mean length of streams of order u.

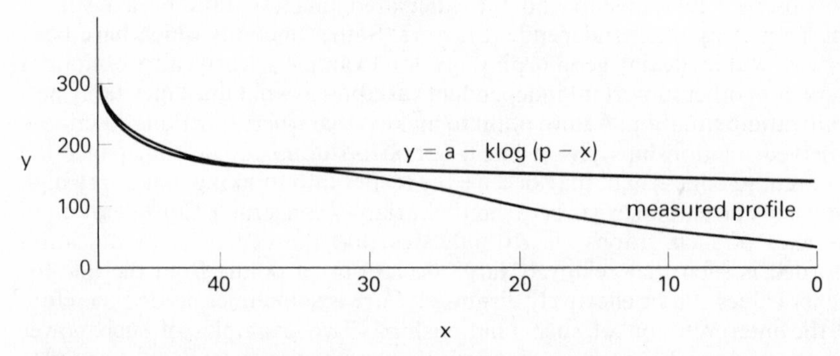

Figure 5.16 Graph of the function $y = a - k \log (p - x)$ applied to an upper section of the River Mole. After Green (1935).

It has been impossible to give more than a very few examples of both linear and non-linear functions encountered in physical geography, but it is hoped that the reader now feels fairly confident in approaching texts or papers where such functions are rife. Other features of mathematics which the reader may encounter in the same literature fall into rather different categories. The next chapter is concerned with the development of one aspect of functions approached but not explained in this chapter—the further treatment of polynomials through *series* and *expansions.*

6
Series and progressions

6.1 Arithmetic and geometric progressions

In two of the previous chapters we have been concerned with functions relating two or more variables. Functions imply *continuous* change in dependent with independent variables. Consider the situation, for example, where the progressive accumulation of sediment in a reservoir is measured purely on an annual basis. The cumulative silting of the reservoir can be represented in table form as a sequence of values which indicate an annual reduction in reservoir storage capacity. Suppose that when construction was complete (year 0) some a cubic metres of sediment had already accumulated, and that for each successive year a further d cubic metres of sediment are deposited in the reservoir. After five years the total *cumulative* sedimentation within that reservoir will be:

$$a+d+d+d+d+d = a+5d$$

after ten years, $a+10d$ and so on. We would then have a *progression* of annual cumulative sedimentation for that reservoir as follows: $a, a+d, a+2d, a+3d, a+4d, a+5d, a+6d, \ldots$ This sequence is known as an *arithmetic* progression of values, with each successive value formed by adding d (termed the *common difference*) to the previous value. If we were to let x represent the year, and y the sedimentation, then a linear function $y = dx+a$ would describe the same sequence for integer changes in x. Thus if $a = 1$ and $d = 2$, then the resultant arithmetic progression of values for sedimentation would be 1, 3, 5, 7, 9, ..., the same as if we were to substitute in for $x = 0, 1, 2, 3, 4, \ldots$ in the function $y = 2x+1$. Either a or d may be negative, so that the progression 10, 9, 8, 7, 6, 5, ... results when $a = 10$ and $d = -1$ (the function $y = 10-x$, for increasing integer values of x from $x = 0$).

Another group of progressions are those which are *geometric* in form. In the above very simple example of reservoir sedimentation, although the actual *rate* of sedimentation was constant from one year to the next, the proportional decrease in reservoir storage from year to year was not. If we again use arbitrary units such that the cumulative sedimentation can be represented by the function $y = 2x+1$, and assume that the total storage of the reservoir at year 0 was 100, then the percentage decrease in storage capacity between successive years is given by 2·02 ($2\times100/99$) during the first year, 2·06 ($2\times100/97$) during the second, and during the third 2·10 and so on. These figures appear in table 6.1. There is an increased percentage change with each succeeding year, so that by the 49th year the percentage decrease will be $2\times100/(100-2\times49-1)$, 200%, and at some point in the 50th year the reservoir

Table 6.1 Sedimentation in a reservoir at an increasing rate but with constant yearly sediment addition.

Years (x)	0	1	2	3	4	5	6
Accumulated sediment (y)	1	3	5	7	9	11	13
Storage capacity ($100-y$)	99	97	95	93	91	89	87
Change in storage	1	2	2	2	2	2	2
% change	−1·01	−2·02	−2·06	−2·10	−2·20	−2·27	−2·33

will have completely silted up. At this point there will have been an overall reduction in capacity of 100%. Thus the arithmetic progression is identical to that of the application of simple interest in economics, where the interest is a function only of the initial capital. The geometric progression is however, analogous to compound interest. If we were to envisage a similar situation involving the sedimentation of a reservoir, but with a *constant* percentage decrease in storage from one year to the next (that is, a function of *existing* capital to continue the economic analogy), then the progression of annual cumulative sedimentation is as follows. Let us say that each year the effective storage in the reservoir is reduced by 25% of the previous year's storage, then the cumulative reduction in storage for the first six years, assuming that storage at the start was again 100 units, is as appears in table 6.2. In more

Table 6.2 Reduction in reservoir storage capacity at a constant rate but with decreasing magnitude of annual storage loss.

Years (x)	0	1	2	3	4	5	6
Remaining storage	100	75	56·25	42·19	31·64	23·73	17·80
Annual reduction	0	25	18·75	14·06	10·53	7·91	5·93

general terms if we let a equal the original capacity of 100 units, and r equal the percentage change in storage every year, then successive terms in the geometric progression for remaining storage are:

$$a,\ a(0{\cdot}75),\ a(0{\cdot}75)(0{\cdot}75),\ a(0{\cdot}75)(0{\cdot}75)(0{\cdot}75),\ \ldots,$$

or:

$$a,\ ar,\ ar^2,\ ar^3,\ ar^4,\ ar^5,\ \ldots$$

Any progression of values which fits with this particular sequence is a geometric progression. In this particular example the remaining storage (y) after time (x) can be expressed as the exponential function $y = 100(0{\cdot}75)^x$. More elementary examples are given here, together with the respective values for a and r, and their equivalent functions of exponential form:

$a = 1,\ r = 2$	1, 2, 4, 8, 16, 32, 64, 128	$y = 2^x$
$a = 1,\ r = 10$	1, 10, 100, 1,000; 10,000	$y = 10^x$
$a = 3,\ r = 3$	3, 9, 27, 81, 243, 729	$y = 3(3)^x = 3^{x+1}$
$a = 1,\ r = \frac{1}{2}$	$1, \frac{1}{2}, \frac{1}{4}, \frac{1}{8}, \frac{1}{16}, \frac{1}{32}$	$y = \frac{1}{2}^x$

The first term in each progression (a) is called the *initial term* and the constant r is the *common ratio*. Note that the nth term in the progression is ar^{n-1}. The reader will note the close similarity in curve form between the functions above and the exponential function $y = e^x$ (figure 5.7).

A further example from physical geography may help to distinguish further between the two types of progression, and tie them in with the corresponding functions. Throughout the field of pedology and sedimentology the ϕ(phi)-scale is used to indicate particle size. The scale is an *ordinal* scale which attempts to relate simple verbal description of particle type, a *nominal* scale: coarse silt, fine sand, coarse clay and so on, to a range of different particle diameters. The relationship between particle description, particle size and the ϕ-value is shown in table 6.3. The use of very simple verbal description is of very limited application, particularly at very small sizes where a slight change in diameter may bring into play a completely different set of physical constraints. Thus whilst the ϕ-scale is basically an arithmetic progression, where $a = 9$ and $d = -1$, so that successive class intervals begin with 9, 8, 7, 6 etc., the corresponding ranges of particle diameter (mm) take the form of a geometric progression whose terms increase whilst the terms of the ϕ-units decrease. For this geometric progression $a = 0{\cdot}00195$ and $r = 2$, so that, again for the lower end of each class, we have 0·00195, 0·0039, 0·0078, 0·0156 etc. Whilst a ϕ-value of 8·0 indicates a diameter of 0·0039 mm, a value of 0·0 indicates 1·0 mm, and a value of −8·0, 256 mm diameter. Higher up the scale in table 6.3 the category sizes enlarge, so interrupting the progressions. However if we were to subdivide the 'pebble' and 'cobble' classes into four and two respectively, we should find that values for d and r in the two progressions were retained: $\phi = -2{\cdot}0$ to $-3{\cdot}0$ would correspond to diameters between 4·0 and 8·0 mm, $\phi = -3{\cdot}0$ to $-4{\cdot}0$ to diameters between 8·0 to 16·0 mm and so on.

The entire sequence between ϕ and diameter may thus be portrayed as a function of ϕ in terms of diameter. Using the above notation we have for the ϕ-scale:

$$\phi = dx + a$$

Table 6.3 ϕ-scale of particle size.

Description	Diameter (mm)	Units
Boulder	256	−8·0
Cobble	256 to 64	−8·0 to −6·0
Pebble	64 to 4	−6·0 to −2·0
Granule	4 to 2	−2·0 to −1·0
Very coarse sand	2 to 1	−1·0 to 0·0
Coarse sand	1 to 0·5	0·0 to 1·0
Medium sand	0·5 to 0·25	1·0 to 2·0
Fine sand	0·25 to 0·125	2·0 to 3·0
Very fine sand	0·125 to 0·0625	3·0 to 4·0
Coarse silt	0·0625 to 0·0312	4·0 to 5·0
Medium silt	0·0312 to 0·0156	5·0 to 6·0
Fine silt	0·0156 to 0·0078	6·0 to 7·0
Very fine silt	0·0078 to 0·0039	7·0 to 8·0
Coarse clay	0·0039 to 0·00195	8·0 to 9·0
Medium clay	0·00195 to 0·00098	9·0 to 10·0

where x represents the change from one particle diameter category to another, thus

$$\phi = 9 - x \tag{6.1}$$

and for particle diameter (D):

$$D = ar^n = 0{\cdot}00195(2)^x \tag{6.2}$$

From this equation we have:

$$\log D = \log 0{\cdot}00195 + x \log 2$$

and
$$x = \frac{\log D - \log 0{\cdot}00195}{\log 2}$$
$$= -3{\cdot}32 \log D - (-2{\cdot}71/0{\cdot}3010)$$
$$\therefore\ x = -3{\cdot}32 \log D + 9{\cdot}0$$

Substituting for x into the equation (6.1) we have:

$$\phi = -3{\cdot}32 \log D$$

This equation provides a continuous function relating the particle diameter to its ϕ-value.

Problems 6.1

Establish whether the following progressions are geometric or arithmetic and find the values of a, r or d as appropriate.

6.1.1 $2{\cdot}50,\ 1{\cdot}25,\ 0{\cdot}00,\ -1{\cdot}25,\ -2{\cdot}50, \ldots$
6.1.2 $2x,\ 3x,\ 4x,\ 5x,\ 6x, \ldots$
6.1.3 $6,\ 36,\ 216,\ 1296,\ 7776, \ldots$
6.1.4 $2{\cdot}1,\ 8{\cdot}4,\ 33{\cdot}6,\ 134{\cdot}4,\ 537{\cdot}6, \ldots$
6.1.5 $-3{\cdot}0,\ -1{\cdot}5,\ -0{\cdot}75,\ -0{\cdot}375,\ -0{\cdot}1875, \ldots$
6.1.6–6.1.10 Calculate the eighth and tenth terms in each of the above.

6.2 Series

The main use of progressions within physical geography is in the general application of their mathematics to more advanced methods, such as integration (chapter 8) and probability through the binomial theorem (later this chapter and in the next chapter). They have only a relatively limited direct application to physical geography itself. The most useful application of them in mathematics is through the addition of each term in a progression to produce a *series*. For example, $a + (a+d) + (a+2d) + \ldots$ is an arithmetic series, and $a + ar + ar^2 + ar^3 + \ldots$ is a geometric series. A series may be either *finite* or *infinite*. For example we might wish to sum *all* the terms in a progression denoted by P_m where m represents the term number. In this case we can represent the series as:

$$\sum_{m=1}^{m=\infty} P_m$$

indicating the *expansion* of

$$P_1 + P_2 + P_3 + P_4 + P_5 + \ \ldots\ + P_n, \text{ where } n = \infty.$$

Such a series is termed an *infinite* series. By definition therefore, a finite series is one in which the expansion takes place within finite limits, for example:

$$\sum_{m=1}^{m=10} P_m = P_1+P_2+P_3+P_4+\ldots+P_{10}$$

and:

$$\sum_{m=12}^{m=20} P_m = P_{12}+P_{13}+P_{14}+\ldots+P_{20}$$

The expansion of a number of geometric series is of particular importance to the establishment of certain mathematical constants, such as e, and trigonometric values, such as sin x, which we shall see later in this chapter. First of all however, we shall look at how the sums of arithmetic and geometric series may be calculated.

6.3 The Sum of an arithmetic series

Following an earlier definition, the expansion of the arithmetic series P_m is given by:

$$\sum_{m=1}^{m=n} P_m = a+(a+d)+(a+2d)+(a+3d)+\ldots+(a+(n-1)d) \qquad (6.3)$$

There are n terms in this expression, and therefore the final term must involve $d(n-1)$. This expansion can therefore be simplified to:

$$\sum_{m=1}^{m=n} P_m = na+d+2d+3d+4d+5d+\ldots+(n-1)d \qquad (6.4)$$

since there are n terms involving a, leaving merely $(n-1)$ terms in d. By reversing the order in which the terms occur we can write this in an alternative form:

$$\sum_{m=1}^{m=n} P_m = na+(n-1)d+(n-2)d+(n-3)d+\ldots+3d+2d+d \qquad (6.5)$$

By adding equation (6.4) to (6.5) we therefore have:

$$2\sum_{m=1}^{m=n} P_m = 2na+d+(n-1)d+2d+(n-2)d+\ldots+(n-1)d+d$$

and thus:

$$2\sum_{m=1}^{m=n} P_m = 2na+nd+nd+nd+nd+\ldots+nd$$

There are $(n-1)$ terms involving nd and therefore:

$$2\sum_{m=1}^{m=n} P_m = 2na+(n-1)nd$$

$$\therefore \sum_{m=1}^{m=n} P_m = n\left(a+\frac{(n-1)d}{2}\right) \qquad (6.6)$$

More usefully however, we can see that by regrouping the terms in a, n and d within the large parentheses we obtain:

$$\sum_{m=1}^{m=n} P_m = n\left(\frac{a+(a+(n-1)d)}{2}\right) \tag{6.7}$$

or that, the sum of all the terms in the finite arithmetic series P_m is simply given by *the number of terms multiplied by the mean of the first and the last terms.* Thus the sum of the first to tenth terms of a series defined by $a = 1$ and $d = 1$ is:

$$10\frac{1+(1+9)1}{2} = \frac{11 \times 10}{2} = 55$$

For a series which may be expanded to infinity, the sum of a finite number of terms is called the *partial* sum. Thus in the above example, we have the partial sum of the series to the tenth term. Similarly we can find the sum of any sequence of terms within an arithmetic series, so that for the same series the partial sum of the series from the eighth to the 15th terms inclusive (i.e. *eight* terms: 8, 9, 10, 11, 12, 13, 14, 15) is:

$$8\frac{(1+7)+(1+14)}{2} = 8 \times 11{\cdot}5 = 92$$

or in more general terms:

$$\sum_{m=b}^{m=c} P_m = (c-b+1)\frac{(a+(b-1)d)+(a+(c-1)d)}{2}$$

Problems 6.3

Calculate the sums of the following arithmetic series within the specified limits:

6.3.1 $a = 2$, $d = 3{\cdot}5$, $m = 1$ to 10
6.3.2 $a = -0{\cdot}2$, $d = -2$, $m = 1$ to 6
6.3.3 $a = 1$, $d = 6$, $m = 5$ to 7
6.3.4 $a = -1$, $d = 0{\cdot}2$, $m = 3$ to 10
6.3.5 $a = 4$, $d = -0{\cdot}2$, $m = 8$ to 14

6.4 The Sum of a geometric series

Other important results stem from the expansion of a geometric series. Using the general expansion for such a series, we have:

$$\sum_{m=1}^{m=n} S_m = a + ar + ar^2 + ar^3 + ar^4 + \ldots + ar^{n-1} \tag{6.8}$$

If we now multiply equation (6.8) throughout by r we arrive at:

$$r\sum_{m=1}^{m=n} S_m = ar + ar^2 + ar^3 + ar^4 + \ldots + ar^n \tag{6.9}$$

and thus, subtracting (6.9) from (6.8) we have:

$$\sum_{m=1}^{m=n} S_m - r\sum_{m=1}^{m=n} S_m = a - ar^n$$

$$\therefore \sum_{m=1}^{m=n} S_m(1-r) = a(1-r^n)$$

$$\therefore \sum_{m=1}^{m=n} S_m = \frac{a(1-r^n)}{1-r} \qquad (6.10)$$

For example, the sum of the first five terms in the geometric series where $a = 0{\cdot}1$ and $r = 2$ is given by:

$$\frac{0{\cdot}1(1-2^5)}{(1-2)} = \frac{0{\cdot}1-3{\cdot}2}{-1} = 3{\cdot}1$$

A complete expansion of such a geometric series to infinity, or the sum to infinity of an arithmetic series is strictly speaking impossible: indeed it can be argued that since there are already an infinite number of terms in the expansion, the *sum* of all terms represents a number which may exceed infinity, since in many such series the value of each term is greater than one. Such a geometric series, which tends to infinity is called *divergent*.

Consider however, the geometric progression where $-1 < r < +1$. In the expansion of such a series, successive terms will *decrease* in value as r is raised to higher and higher powers. As an illustration, if $r = 0{\cdot}5$, $r^2 = 0{\cdot}25$, $r^3 = 0{\cdot}125$, $r^4 = 0{\cdot}0675$ and so on, we can thus say that as n tends to infinity in the succession of terms to ar^n, the value of r^n itself tends to *zero*. Thus in equation (6.10) as r^n tends to zero and $(1-r^n)$ tends to one, the expression for the sum of the series becomes:

$$\sum_{m=1}^{m=n} S_n = \frac{a}{1-r} \quad \text{for } -1 < r < +1 \qquad (6.11)$$

The sum of the entire expansion within these limits of r is no longer dependent upon n, and thus a *finite* value can be ascribed to the expansion. For example if $a = 64$ and $r = 0{\cdot}5$, then the resulting sum of the expansion will be: $64/0{\cdot}5 = 128$. Note however, that using the first ten terms of the progression, the partial sum of the first five is 124, and that of all ten is 127·875. We can arrive at the same value substituting $n = 10$, $a = 64$ and $r = 0{\cdot}5$ in equation (6.10). The first ten terms in the series are:

64, 42, 16, 8, 4, 2, 1, 0·5, 0·25, 0·125

The sum of the entire series will be approached only as we approach an infinite number of terms. It should be clear then that such a series has similar characteristics to the asymptotic curves of certain functions introduced in the previous chapter. The series expansion *approaches* the sum of the series but only actually reaches it when n itself reaches infinity. Such a series is thus said to be *convergent*, and in this case it is convergent to the sum of 128.

The antecedent precipitation index (P_a) is one particularly useful example of a convergent geometric series in physical geography. For such an index the choice of value of the constant r in the series is often arbitrary. All that is required is that the series reflects the decreasing impact of each day's rainfall back through time, so that $0<r<1$. We might arbitrarily decide that the importance of each day's rainfall is half that of the day following it. In the resulting geometric series we have $r = 0{\cdot}5$, so that if P_t represents the precipitation total t days previously, and the total precipitation for a period of ten days is given by:

$$\sum_{t=1}^{t=10} P_t$$

then the weighting applied to the precipitation of each day in the series is in the form:

$$P_a = \sum_{t=1}^{t=10} P_t r^t \quad (0<r<1)$$

where $t = 1$ is the previous day and $t = 10$ is 10 days previous. For $r = 0{\cdot}5$ then:

$$P_a = P_1 + 0{\cdot}5P_2 + 0{\cdot}25P_3 + 0{\cdot}125P_4 + \ldots + 0{\cdot}0009765P_{10}$$

The value of r can be varied according to local conditions. For example, if we are using the index to construct a river discharge model, then for drainage basins whose regime is 'flashy', and the response to rainfall is rapid, a relatively low value might be substituted. In other areas where the component of river discharge contributed by base flow is high, and reaction to precipitation events is relatively slow, a high value for r may be chosen.

A certain similarity between some series and functions has been noted earlier in the chapter. In the context of geometric series however, if we still assume that $-1<r<+1$, then from equation (6.8) we know that:

$$\sum_{m=1}^{m=n} S_m = a + ar + ar^2 + ar^3 + \ldots + ar^{n-1}$$

and also from equation (6.11) that:

$$\sum_{m=1}^{m=n} S_m = \frac{a}{1-r}$$

Thus, if we let $a = 1$, we can obtain an expansion for a function of the form:

$$y = \frac{1}{1-r} = (1-x)^{-1} = 1 + x + x^2 + x^3 + \ldots + x^{n-1} \qquad (6.12)$$

This is developed further in the next section: the binomial theorem.

Problems 6.4

Calculate the sums of the following geometric series to the terms indicated:

6.4.1 $a = 2$, $r = 0{\cdot}2$ to infinity
6.4.2 $a = 0{\cdot}5$, $r = -0{\cdot}5$ to infinity
6.4.3 $a = 7$, $r = 2$ to the 10th term
6.4.4 $0{\cdot}2 + 0{\cdot}6 + 1{\cdot}8 + 5{\cdot}4 + 16{\cdot}2 + \ldots$ to the tenth term
6.4.5 $0{\cdot}8 - 0{\cdot}256 + 0{\cdot}082 - 0{\cdot}0262 + \ldots$ to infinity

6.5 The binomial theorem

In the previous chapter we introduced the concept of polynomials. It should by now be apparent that there is a close similarity between polynomials of the simple form: $a+bx+cx^2+dx^3$ and so on, and the general expansion for a geometric series given in equation (6.8), except that for a geometric series the values a, b, c and d in the polynomial are all equal. It was also noted in chapter 5 that many polynomials can be factorized by the distributive law of multiplication (chapter 1) into the form $(a+b)(c+d)$ and so on. One particularly useful application of such terms occurs when the pairs of values within each set of brackets are the same—that is, when we are interested in the expansion of the expression $(a+x)^n$. Such an expression is termed a *binomial*, and its expansion (the multiplication out of the entire expression n times) is of particular relevance to probability studies (chapter 7). In the context of probability the value of $(a+x)$ is always equal to one, for example, when a represents the probability of a spun coin landing heads and x the probability of it landing tails. In this particular case, with one coin, there is an equal probability of either face appearing after each spin, and thus we can say that $a = 0{\cdot}5$ and $x = 0{\cdot}5$. This is developed further in chapter 7.

For the present we must limit our discussion to investigating the expansion of the binomial $(a+x)^n$. As a simple example, the expansion of the elementary binomial, when $n = 2$, is as follows:

$$(a+x)^2 = a^2+2ax+x^2$$

Similarly the expansions of $(a+x)^3$ and $(a+x)^4$ are:

$$(a+x)^3 = a^3+3a^2x+3ax^2+x^3$$
$$(a+x)^4 = a^4+4a^3x+6a^2x^2+4ax^3+x^4$$

In each case as the value of n increases, so too does the number of terms in the expansion. Note however, that only one term each of a^n and x^n occurs, and that the coefficients of all terms are symmetrical about the central point in each case: 1, 2, 1 for $n = 2$; 1, 3, 3, 1 for $n = 3$; and 1, 4, 6, 4, 1 for $n = 4$. These are known as the *binomial coefficients*. Note also that the total number of terms in each expansion is always equal to $n+1$. If we represent each binomial coefficient by ${}^nC_r$, indicating the rth term out of a total of n terms (this can be depicted alternatively as ${}^n_rC$ or C^n_r), then when $n = 2$, ${}^2C_0 = 1$, ${}^2C_1 = 2$ and ${}^2C_2 = 1$. For $n = 3$, ${}^3C_0 = 1$, ${}^3C_1 = 3$, ${}^3C_2 = 3$ and ${}^3C_3 = 1$.

Just as there is a symmetry within each expansion for the coefficients ${}^nC_r$, there is also a symmetry between the constant terms for successive expansions with increasing n. These assume a geometric as well as an arithmetic symmetry which culminates in *Pascall's triangle*. The use of Pascall's triangle provides one way of solving for values of ${}^nC_r$ when n is very large, rather than laboriously multiplying out the factors. A portion of the triangle for $n = 1$ to $n = 10$ is shown on page 115. Note that the values for the constants for the expansion at $(n+1)$ are always produced by the addition of the two values above them in the triangle for the expansion at n, such that when $n = 6$, the value of the coefficient ${}^6C_2 = {}^5C_1 + {}^5C_2 = 5+10 = 15$.

Although the calculation of the coefficients ${}^nC_r$ using Pascall's triangle is an improvement on the laborious multiplication of terms long-hand, it is still a

$n = 1$	1
$n = 2$	1 2 1
$n = 3$	1 3 3 1
$n = 4$	1 4 6 4 1
$n = 5$	1 5 10 10 5 1
$n = 6$	1 6 15 20 15 6 1
$n = 7$	1 7 21 35 35 21 7 1
$n = 8$	1 8 28 56 70 56 28 8 1
$n = 9$	1 9 36 84 126 126 84 36 9 1
$n = 10$	1 10 45 120 210 252 210 120 45 10 1

relatively inconvenient way of determining their values. Clearly the value of each coefficient is related to its position in the series (i.e. the value of r) and to the value of n. In fact the coefficient ${}^3C_2$ is given by:

$$\frac{3 \times 2}{2 \times 1} = 3$$

and ${}^5C_3$ is given by:

$$\frac{5 \times 4 \times 3}{3 \times 2 \times 1} = \frac{60}{6} = 10$$

and ${}^7C_5$ is given by:

$$\frac{7 \times 6 \times 5 \times 4 \times 3}{5 \times 4 \times 3 \times 2 \times 1} = \frac{7 \times 6}{2} = 21$$

and so on. We can here introduce a further symbol, very commonly used in mathematics and probability statistics, in that:

$$4 \times 3 \times 2 \times 1 = 4! \quad \text{('four factorial')}$$

Using this notation a general form for the value of each coefficient emerges as:

$$^nC_r = \frac{n(n-1)(n-2)\dots(n-r+1)}{r!}$$

$$= \frac{n!}{r!(n-r)!} \tag{6.13}$$

If we now apply the formula to the coefficients of the expansion $(a+x)^4$ we arrive at:

$$^4C_0 = \frac{4!}{0!(4!)} \quad {}^4C_1 = \frac{4!}{1!(3!)} \quad {}^4C_2 = \frac{4!}{2!(2!)}$$

$$^4C_3 = \frac{4!}{3!(1!)} \quad {}^4C_4 = \frac{4!}{4!(0!)}$$

By convention $0! = 1 = 1!$ and therefore the calculated coefficients using this method for ${}^4C_r$ emerge as: 1, 4, 6, 4, 1.

Finally, the expansion of the binomial series entails the sequence of powers of each of the values within the brackets, which follow the form:

$$a^n,\ a^{n-1}x,\ a^{n-2}x^2,\ a^{n-3}x^3,\ a^{n-4}x^4,\ \ldots,\ a^{n-r}x^r,\ \ldots,\ x^n$$

For example, ignoring the binomial coefficients, the expansion of $(a+x)^5$ involves terms in a^5, a^4x^1, a^3x^2, a^2x^3, a^1x^4, x^5. Thus the complete binomial expansion can be written:

$$(a+x)^n = a^n + \frac{n!}{(n-1)!}a^{n-1}x + \frac{n!}{2!(n-2)!}a^{n-2}x^2 + \ldots + \frac{n!}{r!(n-r)!}a^{n-r}x^r + \ldots + x^n \qquad (6.14)$$

This formula is generally referred to as the *binomial theorem*, and examples of its use in probability can be found in chapter 7.

The binomial expansion can be further developed to provide two expansions for which we shall find a use when attempting certain aspects of integration in chapter 8. If we were to write the expansion for the binomial theorem (equation (6.14)) in an alternative and more easily managed form:

$$(a+x)^n = a^n + na^{n-1}x + \frac{n(n-1)}{2!}a^{n-2}x^2 + \frac{n(n-1)(n-2)}{3!}a^{n-3}x^3 + \ldots + \frac{n(n-1)(n-2)\ldots(n-r+1)}{r!}a^{n-r}x^r + \ldots + x^n \qquad (6.15)$$

then, dividing through by a we have:

$$a^n\left(1+\frac{x}{a}\right)^n = a^n\left[1 + n\left(\frac{x}{a}\right) + \frac{n(n-1)}{2!}\left(\frac{x}{a}\right)^2 + \frac{n(n-1)(n-2)}{3!}\left(\frac{x}{a}\right)^3 + \ldots + \frac{n(n-1)(n-2)\ldots(n-r+1)}{r!}\left(\frac{x}{a}\right)^r + \ldots + \left(\frac{x}{a}\right)^n\right]$$

If we now take $n = -1$, for the expansion itself, that is, for the expansion

$$\left(1+\frac{x}{a}\right)^{-1} \quad \text{or} \quad 1\bigg/\left(1+\frac{x}{a}\right)$$

then we have:

$$\left(1+\frac{x}{a}\right)^{-1} = 1 + -1\left(\frac{x}{a}\right) + \frac{-1.-2}{2}\left(\frac{x}{a}\right)^2 + \frac{-1.-2.-3}{3.2}\left(\frac{x}{a}\right)^3 + \ldots$$

$$= 1 - \frac{x}{a} + \left(\frac{x}{a}\right)^2 - \left(\frac{x}{a}\right)^3 + \left(\frac{x}{a}\right)^4 - \ldots + (-1)^r\left(\frac{x}{a}\right)^r \qquad (6.16)$$

Thus the expansion includes coefficients whose values are always either $+1$ or -1, and which alternate through the expansion. If the value of x/a is greater than or equal to one, then the series is divergent. If on the other hand x/a lies between zero and one, then the series is convergent. This particular type of series is known as an *oscillating* series if it is convergent, since the step-wise addition of individual terms produces alternately positive and negative steps about the convergent sum. The reader will already have encountered an oscillating series in problem 6.4.5. Note however, that if x/a lies between -1 and zero, then the expansion ceases to oscillate, and becomes:

$$\left(1-\frac{x}{a}\right)^{-1} = 1+\frac{x}{a}+\left(\frac{x}{a}\right)^2+\left(\frac{x}{a}\right)^3+\ldots+\left(\frac{x}{a}\right)^r+\ldots \tag{6.17}$$

Problems 6.5

Obtain expansions of the following expressions using the binomial theorem:

6.5.1 $(1+x)^5$
6.5.2 $(3-y)^{-2}$ to the 5th term
6.5.3 $(5y+2x)^4$
6.5.4 $(0{\cdot}5+x)^{-1}$ to the 5th term
6.5.5 $(x+2)^4$

6.6 Further Examples of series

We can further use the binomial theorem to define certain mathematical functions, which we met in chapters 3 and 5. A further development of equation (6.15) is in the derivation of the mathematical constant e. The constant is obtained by the expansion of the expression

$$\left(1+\frac{1}{n}\right)^n$$

as n tends to infinity.

It is therefore only possible to approximate its value by calculating the partial sum of a manageable number of terms, as the series is convergent. However, we can simply calculate it to four decimal places; the precision to which it is commonly given in tables and used in calculation. The expansion can be obtained by substituting $a = 1$ and $x = 1/n$ in the binomial expansion given in equation (6.15). Thus,

$$\left(1+\frac{1}{n}\right)^n = 1+n.1^{n-1}\frac{1}{n}+\frac{n(n-1)}{2!}1^{n-2}\left(\frac{1}{n}\right)^2+\frac{n(n-1)(n-2)}{3!}1^{n-3}\left(\frac{1}{n}\right)^3+\ldots+\frac{1}{n^n}$$

$$= 1+1+\frac{n}{n}\frac{(n-1)}{n}\frac{1}{2!}+\frac{n}{n}\frac{(n-1)}{n}\frac{(n-2)}{n}\frac{1}{3!}+\ldots+\frac{1}{n^n}$$

As n tends to infinity then the fractions $(n-1)/n$ and $(n-2)/n$ for example, tend to one, and thus the expansion for e is:

$$\begin{aligned} e &= 1+1+\frac{1}{2!}+\frac{1}{3!}+\frac{1}{4!}+\ldots \\ &= 1+1+0{\cdot}5+0{\cdot}16667+0{\cdot}04167+0{\cdot}00833 \\ &\quad +0{\cdot}00139+0{\cdot}00020+0{\cdot}00003 \end{aligned} \tag{6.18}$$

to the 8th term, and thus:

$$e = 2{\cdot}71829 = 2{\cdot}7183 \text{ to four decimal places}$$

The addition of further terms makes very little difference to the calculated value for the constant. The value of e is irrational and the series converges to the true value only at infinity.

By a similar process we can expand the function e^x to produce the *exponential series*:

$$e^x = 1+x+\frac{x^2}{2!}+\frac{x^3}{3!}+\frac{x^4}{4!}+\ldots+\frac{x^r}{r!}+\ldots \quad (6.19)$$

or e^{-x}, which produces in expansion the oscillating series:

$$e^{-x} = 1-x+\frac{x^2}{2!}-\frac{x^3}{3!}+\frac{x^4}{4!}-\ldots+\frac{x^r}{r!}+\ldots \quad (6.20)$$

These series are examples of *Maclaurin's series.* Maclaurin showed that where a series can be expressed as a power series and is differentiable, then, if the sum of the series is $S = f(x)$:

$$f(x) = f(0)+\frac{f'(0)}{1!}x+\frac{f''(0)}{2!}x^2+\frac{f^{(3)}(0)}{3!}x^3+\ldots \quad (6.21)$$

using the notation adopted in chapter 5 for derivatives. Thus, for $f(x) = e^x$, $f'(x) = e^x$ and so on, so that:

$$e^x = e^0+xe^0+x^2\frac{e^0}{2!}+x^3\frac{e^0}{3!}+\ldots$$

$$= 1+x+\frac{x^2}{2!}+\frac{x^3}{3!}+\ldots$$

In a similar way we may use the expansion of Maclaurin's series to show that the sine of any angle, measured in radians, is given by the expression:

$$\sin x = x-\frac{x^3}{3!}+\frac{x^5}{5!}-\frac{x^7}{7!}+\ldots \quad (6.22)$$

For example, if x = 30 degrees = 0·5236 radians, then:

$$\sin x = 0{\cdot}5236-\frac{0{\cdot}5236^3}{6}+\frac{0{\cdot}5236^5}{120}-\ldots$$

$$= 0{\cdot}5236-0{\cdot}0240+0{\cdot}0003\ldots$$
$$= 0{\cdot}4\dot{9} \quad \text{(the actual value is } 0{\cdot}5\dot{0}\text{)}$$

Again this is an oscillating series, but it is important to note one useful property of this expansion; namely, that for sine values when x is very small, as $x \to 0$, then $\sin x \to x$, since only the first term in the expansion is significant at this magnitude of angle (equation (6.22)). Thus the sine of any angle less than about 0·1 radians (approximately six degrees) is very closely approximated to four decimal places by the radian measure itself, since for x = 0·1, the second term in the expansion subtracts only 0·0002 from the value. The expansion for $\cos x$ can of course be obtained by differentiating the series with respect to x, or substituting into equation (6.21), so that:

$$\cos x = 1 - \frac{3x^2}{3!} + \frac{5x^4}{5!} - \frac{7x^6}{7!} + \ldots$$

$$= 1 - \frac{x^2}{2!} + \frac{x^4}{4!} - \frac{x^6}{6!} + \ldots \tag{6.24}$$

Here we must note that the derivative of a series is the derivative of each of its individual terms. Other trigonometric functions can be determined as expansions but the individual terms of the expansion become more complex. So, we can give the expansion for tan x only within the limits $-\pi/2 < x < +\pi/2$ as follows:

$$\tan x = x + \frac{x^3}{3} + \frac{2x^5}{15} + \frac{17x^7}{315} + \ldots \tag{6.25}$$

There are a number of other functions for which the binomial theorem may be used to obtain specific values. In particular we can develop expansions for deriving logarithmic values. By using these we can do away, if we wish, with the need for tables in their calculation. Although the proof is beyond the scope of this book it is nevertheless useful to remember that the natural logarithm of certain numbers can be obtained by using the expansion for $\log_e(1+x)$, giving the oscillating series:

$$\log_e(1+x) = x - \frac{x^2}{2} + \frac{x^3}{3} - \frac{x^4}{4} + \ldots \tag{6.26}$$

This expression is valid only for $-1 < x < +1$ and may again be derived from Maclaurin's series. The natural logarithm can thus be obtained for any value in the range zero to $1{\cdot}\dot{9}$. For example, putting $x = -0{\cdot}4$ we can write

$$\log_e 0{\cdot}6 = \log_e (1{\cdot}0 - 0{\cdot}4)$$

$$= -0{\cdot}4 - \frac{0{\cdot}16}{2} - \frac{0{\cdot}064}{3} - \frac{0{\cdot}00256}{4} - \ldots$$

$$= -0{\cdot}5056 = \bar{1}{\cdot}4944$$

Similarly, putting $x = 0{\cdot}2$

$$\log_e 1{\cdot}2 = \log_e (1{\cdot}0 + 0{\cdot}2)$$

$$= 0{\cdot}2 - \frac{0{\cdot}04}{2} + \frac{0{\cdot}008}{3} - \frac{0{\cdot}0016}{4}$$

$$= 0{\cdot}1823$$

This and many other expansions are possible by utilizing basic definitions of each function in combination. Thus the two hyperbolic functions (equations (5.26) and (5.27)) may be calculated by using their definitions in terms of e^x and e^{-x}. Thus:

$$\sinh x = \frac{e^x - e^{-x}}{2}$$

$$= \tfrac{1}{2}\left(1 + x + \frac{x^2}{2!} + \frac{x^3}{3!} + \ldots\right) - \tfrac{1}{2}\left(1 - x + \frac{x^2}{2!} - \frac{x^3}{3!} + \ldots\right)$$

$$= \tfrac{1}{2}\left(2x + \frac{2x^3}{3!} + \frac{2x^5}{5!} + \ldots\right)$$

$$= x + \frac{x^3}{3!} + \frac{x^4}{5!} + \ldots \qquad (6.27)$$

Similarly,

$$\cosh x \qquad = \frac{e^x - e^{-x}}{2}$$

$$= 1 + \frac{x^2}{2!} + \frac{x^4}{4!} + \frac{x^6}{6!} + \ldots \qquad (6.28)$$

More complex forms can be solved similarly. For example, the expansion of $e^{\sin x}$ could be obtained from the combination of the two expansions e^x and sin x. The calculation for this can be lengthy but the resultant expansion for $e^{\sin x}$ is:

$$e^{\sin x} = 1 + x\frac{x^2}{2!} - \frac{3x^4}{4!} + \frac{8x^5}{6!} - \ldots \qquad (6.29)$$

Problems 6.6

Using the expansions given in section 6.6 obtain values for the following and check your answer with reference to the relevant tables:

6.6.1 sin 56° 02′
6.6.2 $\log_e$ 1·7240
6.6.3 tan 0·7854 radians
6.6.4 e^3
6.6.5 $e^{\sin x}$ (solution is equation (6.27))

7

Probability

A brief introduction to some of the mathematics involved in statistics has been provided by the consideration of the binomial series in the previous chapter (section 6.5). Probability is one of the more important applications of mathematics to physical geography. A knowledge of the mathematics of probability is essential to an understanding of many parametric statistical techniques, and we have now covered sufficient mathematical ground in this book to permit a basic understanding of probability. There are many occasions when the analysis of data from physical geography reveals that the frquency of occurrence of particular events follows known probability laws, the derivation of which lies within mathematics rather than within statistics. In this chapter we discuss the mathematics behind probability and probability distributions, and some of the applications of probability in physical geography. The applications are twofold: firstly, in the statistical sense when assessing the significance of conclusions arrived at using statistical techniques, and secondly, in the direct sense in estimating for example, the likelihood of occurrence of a severe drought or flood. We have already introduced the application of probability to statistical methods in chapter 5 and a summary of basic statistical tests is to be found in appendix 5. These aspects of probability are developed further by Davidson (1978).

In the direct application of probability to physical geography two distinct types of probability can be specified. Firstly, there is the *empirically* derived probability based upon measurement through instrumentation—for example, where empirical probabilities are derived for particular river discharges from long-term field measurement. Secondly, theoretical or *a priori* probabilities can often be postulated—for example, for every spin of an unbiased coin there is an equal chance of it landing heads or tails. Thus we are able to say that for such a coin there is a 50% chance or a probability of 0·5 that it will land heads after being spun, and an equal chance that it will land tails. A similar criterion can be advanced for the movement of river bedload. Bedload movement will occur once velocity exceeds a critical value determined by the bedload size distribution for the river. Let us take a population of rivers in an hypothetical geographical region. At any one instant some rivers will have static bedloads and others will have mobile bedloads. For convenience and clarity we shall assume that bedloads are either static or mobile for the entire length of any river. The two conditions are *mutually exclusive* since rivers either have or have not a static bedload. Under certain unique conditions it is possible to conceive of exactly 50% of rivers having mobile bedloads, and similarly exactly 50% having static bedloads. However, for the majority of time the division between rivers with mobile and static bedloads will be unequal, and this *bias* will influence the chances of our finding a river of each type, taking any river at random from the area.

7.1 Probability space

These and further probability concepts can be developed using sets delimiting 'probability space' depicted in the form of what are called *Venn diagrams*. In the most simple case we can illustrate an elementary presence/absence or black/white situation as a set S defined as containing all rivers in the above area. The set S itself is not to be confused with the actual geographical area however. The set is a totally abstract concept. Within this set we must imagine two mutually exclusive subsets, one for static and one for mobile bedload, whose areas are proportional to the probability of each occurring (figure 7.1). Since set S represents all rivers then clearly its area is equivalent to the probability $P(S)$ of 1·0, so that if the subsets are labelled A and $\sim A$ ('not A') then:

$$P(A)+P(\sim A) = 1{\cdot}0 \tag{7.1}$$

and:

$$P(A) = M(A)/M(S) \tag{7.2}$$

where $M(A)$ is the area of subset A and $M(S)$ is the area of subset S. The probability $P(\sim A)$ can be similarly represented. We have now defined probability in terms of the relative areas of two subsets containing mutually exclusive cases.

There are however, in addition, a number of characteristics of rivers which are not necessarily mutually exclusive. For example, a river may be either polluted or not according to clearly defined criteria, but polluted rivers can also be either ephemeral or perennial, although we naturally hope there are few with the first combination! Within set S we can now introduce two further subsets: B (polluted) and C (ephemeral). The area within set S not contained within either is proportional to the probability of a river being neither ephemeral nor polluted (figure 7.2) or, $P(\sim B \cap \sim C)$—the symbol $\cap$ is used in set theory to indicate the conjunctive 'and' or 'intersection', whilst $\cup$ indicates the disjunctive 'or' or 'totality'. The area of overlap thus indicates the probability of rivers being *both* ephemeral and polluted, or $P(B \cap C)$ or $P(BC)$.

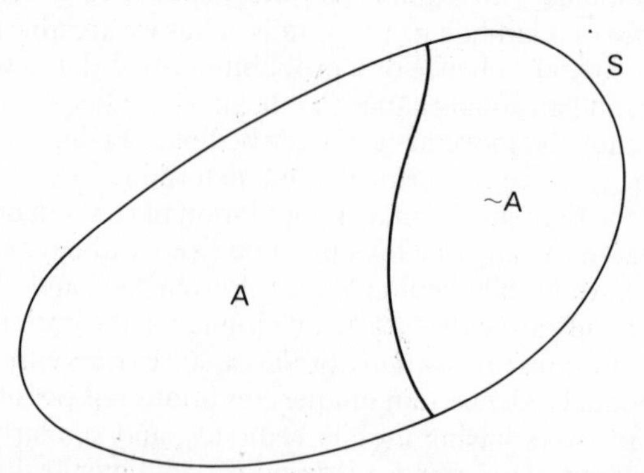

Figure 7.1 Probability space: mutually exclusive subsets.

The total shaded area represents the probability of occurrence for rivers which are *either* polluted or ephemeral. Now we must consider the computation of the probabilities of these different combinations. Clearly the total probability embracing all cases must be 1·0 or 100%, the area depicted by set S. If there were no overlap between subset B and subset C then the probability of either one or the other occurring would be $P(B)+P(C)$, in turn proportional to the area measures $M(B)$ and $M(C)$. This computation for the example shown in figure 7.2 however, would include the area $M(BC)$ twice, and thus the expression for the probability $P(B \cap C)$ is given as:

$$P(B \cap C) = \frac{M(B)+M(C)-M(BC)}{M(S)} = P(B)+P(C)-P(BC) \qquad (7.3)$$

This consideration of probabilities which are not mutually exclusive serves also to introduce *conditional* probability. The area of overlap between subsets B and C in figure 7.2 indicates that a small proportion of rivers are both polluted and ephemeral. The probabilities of a polluted river (B) being ephemeral (C), or and ephemeral river (C) being polluted (B) can be represented respectively as $P(C|B)$ and $P(B|C)$. These are conditional probabilities: the probability of C if B and B if C. Reference to figure 7.2 will

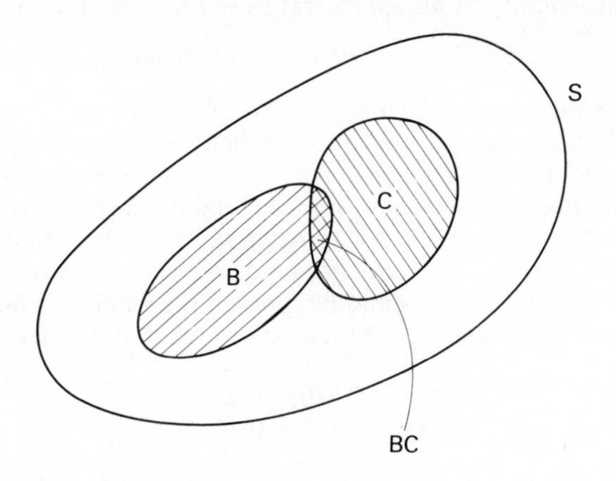

Figure 7.2 Probability space: overlapping subsets.

demonstrate that we are no longer concerned exclusively with the set S, but with the relative areas of subsets B and C, and in particular with their overlap BC. So, for the general computation of $P(B|C)$ we have:

$$P(B|C) = \frac{M(BC)}{M(C)} = \frac{M(BC)/M(S)}{M(C)/M(S)} = \frac{P(BC)}{P(B)} \qquad (7.4)$$

and for $P(C|B)$:

$$P(C|B) = \frac{M(BC)}{M(B)} = \frac{M(BC)/M(S)}{M(B)/M(S)} = \frac{P(BC)}{P(B)} \qquad (7.5)$$

These are further developed in *Bayes's Theorem*. By multiplying both sides of equation (7.5) by $P(B)$ we have:

$$P(C|B)P(B) = P(BC)$$

and substituting in for $P(BC)$ in equation (7.4) we arrive at:

$$P(B|C) = \frac{P(C|B)P(B)}{P(C)} \tag{7.6}$$

which is Bayes's theorem combining both conditional probabilities $P(B|C)$ and $P(C|B)$.

These conditional probabilities may be illustrated with respect to a worked example. Let us assume that the probability of a river being polluted $P(B)$ is 0·4, that the probability of a river being ephemeral $P(C)$ is 0·1, and also that the probability of a river being both polluted and ephemeral $P(BC)$ is 0·05. Clearly then the probability of an ephemeral river being polluted is 0·5 since exactly half the area in subset BC lies within subset C. Using equation (7.4) above to check, we have:

$$P(B|C) = 0{\cdot}05/0{\cdot}1 = 0{\cdot}5$$

Similarly the probability of an ephemeral river being polluted is given by:

$$P(C|B) = 0{\cdot}05/0{\cdot}4 = 0{\cdot}125 \text{ (from (7.5))}$$

We can take a further example to illustrate the different probabilities. Suppose we are particularly concerned during the holiday month of July, about the chances of hot, sticky weather which maybe the pet dog does not like! Our data are somewhat limited since we know only the total number of July days in the last ten years when it has been hot (greater say than 25 °C) and the total number of days during which the mean dew point has exceeded 18 °C, symptomatic of a 'close' day. If A and B represent the number of hot days and humid days respectively and $A = 73$ days and $B = 50$ days then,

$$P(A) = 73/310 = 0{\cdot}236$$
$$P(B) = 50/310 = 0{\cdot}161$$

The probability of any day being *both* warm and humid, $P(AB)$, is not the simple arithmetic sum of these probabilities since some of A will have been humid and some not humid. The two subsets A and B may overlap. Suppose also that as a result of our researches we are able to say that only 14 humid days were warm as well, then:

$$P(A|B) = 14/50 = 0{\cdot}280$$

Now we may use equation (7.4) to say that:

$$P(A|B) = P(AB)/P(B)$$

and

$$P(AB) = P(A|B)P(B)$$
$$= 0{\cdot}280 \times 0{\cdot}161 = 0{\cdot}045$$

Thus there is only a 0·045 (4·5%) chance of any one day in July being *both* warm and humid. We can also say, using Bayes's theorem, that the probability of a warm day also being humid is:

$$P(B|A) = (0{\cdot}280 \times 0{\cdot}161)/0{\cdot}236$$
$$= 0{\cdot}352 \text{ or } 35{\cdot}2\%$$

Let us now consider a completely different set of probabilities. Suppose that the probability of an ephemeral river being polluted is the same as for *any* river being polluted. Here we are saying that if a river is ephemeral this will make no difference to our expectation of it being polluted; or $P(B|C) = P(B)$. In this case we can say that the two subsets B and C are *independent* of one another, a term which has been developed in a similar respect in chapter 3. In these circumstances the probability of both occurring at the same time is given by the product of each individual probability, since:

$$P(B|C) = P(BC)/P(C)$$

and $$P(B) = P(B|C) \text{ if they are independent,}$$

and thus: $$P(B) = P(BC)/P(C)$$

so that: $$P(BC) = P(B)P(C) \tag{7.7}$$

7.2 Chain probabilities

Thus far we have been concerned with probabilities whose outcomes have been independent of previous events. We have considered the distinction between independent and dependent probabilities. However, situations arise which involve some degree of dependence between the outcome of a previous event and that of its successor. For example, if we imagine a bag containing marbles of different colours the removal of each marble inevitably reduces the total number of marbles of that colour in the bag, correspondingly reducing the probability of obtaining a marble of that colour at the next and succeeding draws. The ultimate outcome after say ten draws, depends upon an accumulation of the outcomes of the previous draws. Our treatment of probability so far has not considered the 'chain' probabilities so created. To be strictly accurate we are now in a position to draw a distinction between *probabilistic* studies and *stochastic* processes. Probabilistic processes are entirely time-independent, but stochastic processes imply that the order in which events occur is important in determining the final outcome. Within physical geography there also exists a group of processes where a certain degree of time dependence is inferred. The most well known of these is the simulation process which generates random walks to imitate the evolution of stream networks. In the construction of a random walk, the direction in which a particular step proceeds is determined by the generation of a sequence of random numbers, with the constraint when simulating a river network that the next step cannot be 'uphill'. Clearly we have here a simple system of chain probabilities, with the spatial outcome of the previous generation limiting the outcome of the next generation. For a more detailed discussion of random walks as such the reader is referred to a standard text in fluvial geomorphology, such as Leopold, Wolman and Miller (1964) or Gregory and Walling (1973). The generation of random walks is one of the more elementary chain processes. By further limiting the example such that only two outcomes are possible at each step in the chain, and by ascribing probability values to each outcome, we can define *Markov chains*. These have been applied to the movement of scree slopes (Thornes 1971) and to the occurrence of daily rainfall, regardless of magnitude (Gabriel and Neumann 1962). Using daily rainfall data for a site in Wales, we shall now illustrate some of the various properties of persistence and periodicity of daily rainfall which can be represented by Markov chains as derived initially by Gabriel and Neumann.

The chain is initially determined by two conditional probabilities p_1 and p_0, where:

$$p_1 = P(\text{wet day previous day } \textit{wet})$$

and

$$p_0 = P(\text{wet day previous day } \textit{dry})$$

The probability, p_0, is known as the *transitional* probability; the probability of change from dry to wet on consecutive days. The two probabilities can easily be determined from a set of data. Suppose that for our site analysis of daily rainfall has revealed that out of a sample of 1,280 consecutive days, rainfall occurred on 728 of them. Clearly we can obtain empirically the overall probability of a day on which some rain occurs (P) as:

$$P = 728/1280 = 0{\cdot}57 \tag{7.8}$$

In a similar way the conditional probabilities p_1 and p_0 can be obtained:

$$p_1 = \frac{\text{total no. of wet days following a wet day}}{\text{total no. of wet days}}$$

$$p_0 = \frac{\text{total no. of wet days following a dry day}}{\text{total no. of dry days}}$$

A total of 552 wet days were preceded by a wet day, 176 wet days followed a dry day, and there were 552 dry days (by coincidence). The empirically derived probabilities are therefore:

$$p_1 = 552/728 = 0{\cdot}76 \tag{7.9}$$

$$p_0 = 176/552 = 0{\cdot}32 \tag{7.10}$$

Already it must be clear that some quite sophisticated developments can be made; some degree of persistence of wet days has emerged. Note in addition that the sum of these two probabilities is greater than 1·0, they are not independent. By definition the probability of a dry day following a wet day is $1 - p_1 = 0{\cdot}24$ and similarly the probability of a dry day following another dry day is 0·68. We are now in a position to construct a literal chain of probabilities of wet or dry days greater than one day in advance. This chain is drawn in figure 7.3. The probability of a wet day following a wet day (or $P(W|W)$) is given by p_1; that of a dry day following a wet day ($P(D|W)$) by $1-p_1$, and so on. So the probability of a wet day two days after a wet day is given by:

$$\begin{aligned} P_w &= P(W|W|W) + P(W|D|W) \\ &= p_1 . p_1 + p_0(1-p_1) \\ &= 0{\cdot}76 \times 0{\cdot}76 + 0{\cdot}24 \times 0{\cdot}32 = 0{\cdot}64 \end{aligned}$$

Similarly the probability of a wet day two days after a dry day is:

$$P_d = P(W|W|D) + P(W|D|D)$$
$$= p_1 \cdot p_0 + p_0(1 - p_0)$$
$$= 0{\cdot}76 \times 0{\cdot}32 + 0{\cdot}32 \times 0{\cdot}68 = 0{\cdot}46$$

The computation of subsequent probabilities in the chain by this method becomes a lengthy and tedious process, and the calculation is more easily

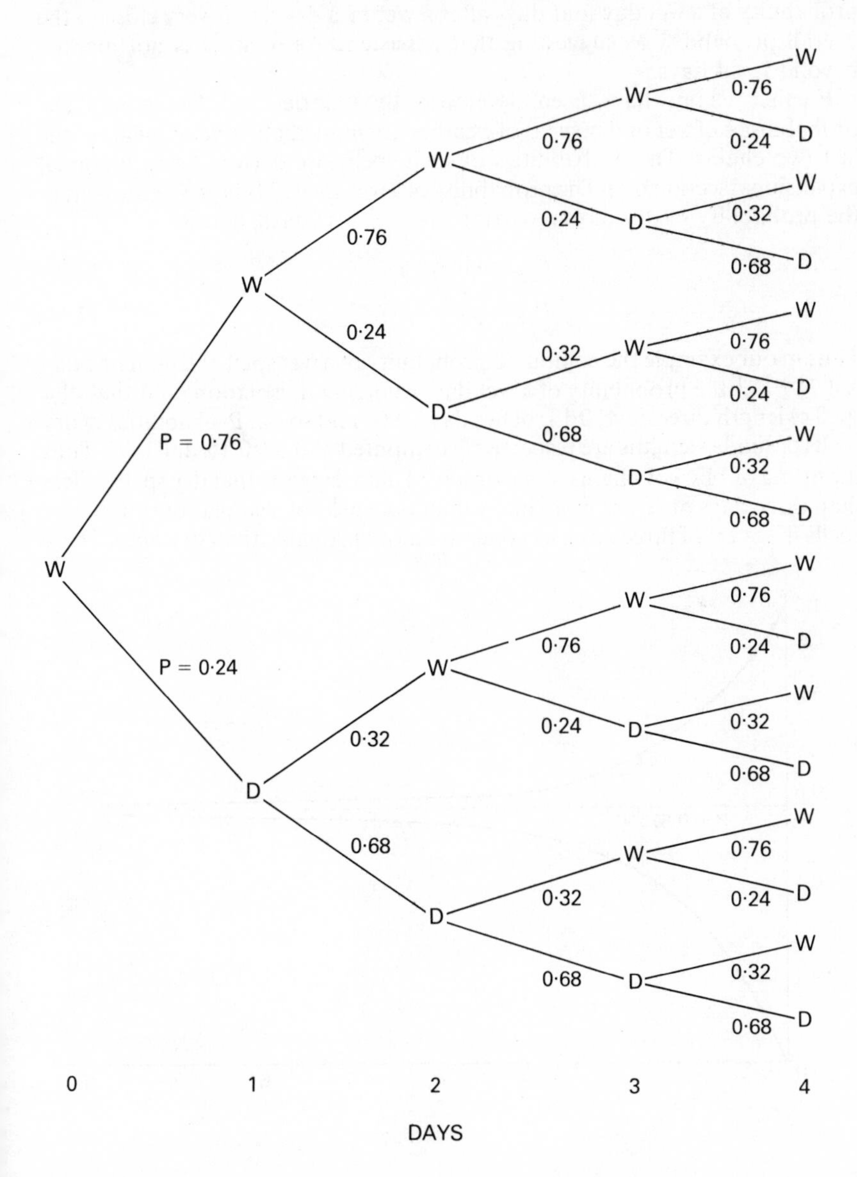

Figure 7.3 Diagrammatic portrayal of Markov chain probabilities for $i = 1$ to 4 days.

made using the following two equations derived for P_w and P_d:

$$P_w = P + (1-P)d^i \tag{7.11}$$

$$P_d = P - Pd^i \tag{7.12}$$

where P is the overall probability of a wet day, and $d = p_1 - p_0$. In both cases the curves of probabilities i days after a wet or a dry day are asymptotic to the probability P. These are shown in figure 7.4. In each case for the data used, the probability of a wet day four days after a wet or a dry day is very close to the overall probability P, suggesting that persistence of weather is not marked beyond $i = 4$ days.

Further chains have been developed by Gabriel and Neumann. The probabilities of wet or dry *spells* of weather are immediate developments of the last two chains. The probabilities of such spells are derived using binomial expansions (section 6.5). The probability of a wet spell of k days is equivalent to the probability of dry days recurring every $(k+1)$ days, and is:

$$P_{sw} = (1-p_1)p_1{}^{k-1} \tag{7.13}$$

$$P_{sd} = p_0(1-p_0)^{k-1} \tag{7.14}$$

Thus in our example the computed probability of a wet spell of length one day is $0{\cdot}24\ P_1^{1-1}$ the probability of a wet day occurring in isolation; and that of a spell of length three days, 0·14; of five days, 0·08 and so on. Probabilities of dry spells of similar lengths are respectively computed as 0·32, 0·15, and 0·07. Thus the fitting of Markov chains to the observed data suggests that dry spells of less than three days or so are more likely than wet spells of a similar duration. For spells in excess of three days, wet ones are more probable than dry ones. These

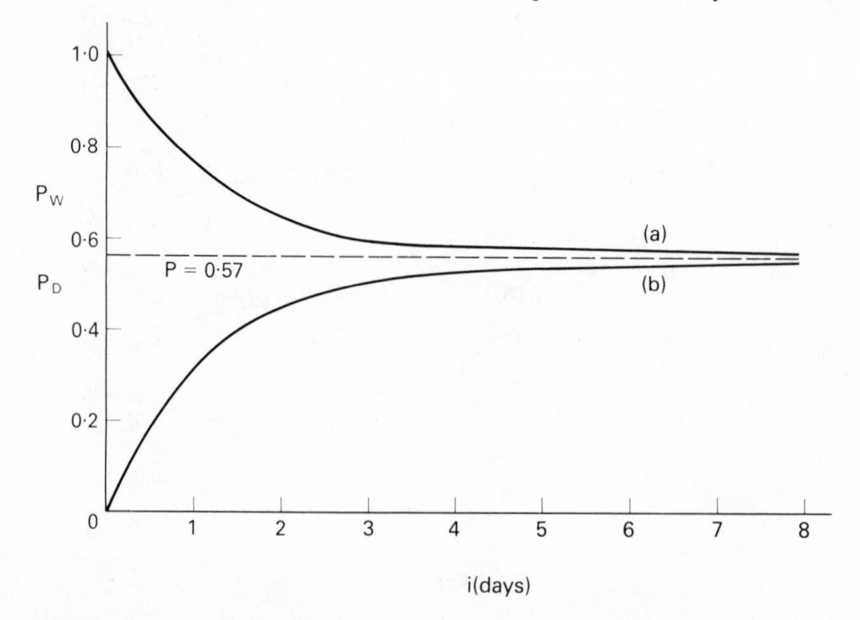

Figure 7.4 Markov probabilities of : (a) a wet day *i* days after a wet day; and (b) a wet day *i* days after a dry day.

chains can be developed further to estimate the probability of *successive* spells of dry and wet days—an index of periodicity. Gabriel and Neumann compute the probability of successive wet and dry spells of n days each (periodicity of $2n$ days) as:

$$P_p = \frac{(1-p_0)^{n-1} - p_1^{n-1}}{(1-p_0-p_1)} p_0(1-p_1) \tag{7.15}$$

The substitution of $p_1 = 0{\cdot}76$ and $p_0 = 0{\cdot}32$ in this expression yields notably low probabilities of periods of all lengths, a peak probability of 0·13 occurring for $n = 4$ days.

As illustrated with the fitting of curves in chapter 5 the approach offered by Markov chains inevitably generalizes the true stochastic sequence involved. The chains derived should be checked against probabilities of specific meteorological events as directly derived from the data. As with all attempts at fitting theoretical statistical models, it is rare for an exact correspondence between observation and theory to occur. With Markov chains only the initial conditional probabilities (p_1 and p_0) and the overall probability P, are derived directly from the data. All other derived probabilities are based on the assumption of dependence of conditions upon those of the previous day. The goodness of fit of the calculated chains of probability can only be seen in the light of the calculation of probabilities directly from the data and the statistical comparison of the two. Where consistent agreement is found using data from different sources and areas then a tacit assumption can be made that daily rainfall follows a Markov chain model. Gabriel and Neumann in their treatment of daily rainfall at Tel Aviv, Israel, found a close parity between observed frequency and computed Markov probability.

7.3 Matrix solution of Markov probabilities

One quite important use of matrices lies with their application in the field of probability studies, and this is shown here in connection with Markov probabilities. If we rewrite the four transitional probabilities indicating the four possible changes from wet or dry on successive days as $P_{ww} = 0{\cdot}76$, $P_{dw} = 0{\cdot}32$, $P_{wd} = 0{\cdot}24$ and $P_{dd} = 0{\cdot}68$ (ww indicates wet day to wet day and so on), then we may write in matrix form:

$$\text{(tomorrow)} \quad \begin{matrix} & \text{w} & \text{d} \\ \text{w} & \left(\begin{matrix} 0{\cdot}76 \\ 0{\cdot}24 \end{matrix}\right. & \left.\begin{matrix} 0{\cdot}32 \\ 0{\cdot}68 \end{matrix}\right) \\ \text{d} & & \end{matrix} \quad \text{(today)} \tag{7.16}$$

If we now represent the state (i.e. whether it was wet or dry) of days before and days after by means of a diagonal matrix,

$$\begin{matrix} & \text{w} & \text{d} \\ \text{w} & 1 & 0 \\ \text{d} & 0 & 1 \end{matrix}$$

the probability of a wet day or a dry day n days after any wet or dry day can be summarized in matrix form as:

$$\begin{pmatrix} 0{\cdot}76 & 0{\cdot}32 \\ 0{\cdot}24 & 0{\cdot}68 \end{pmatrix}^n \begin{pmatrix} 1 & 0 \\ 0 & 1 \end{pmatrix} \tag{7.17}$$

So for $n = 2$, we have:

$$P = \begin{pmatrix} 0{\cdot}76 & 0{\cdot}32 \\ 0{\cdot}24 & 0{\cdot}68 \end{pmatrix} \begin{pmatrix} 0{\cdot}76 & 0{\cdot}32 \\ 0{\cdot}24 & 0{\cdot}68 \end{pmatrix} \begin{pmatrix} 1 & 0 \\ 0 & 1 \end{pmatrix}$$

$$= \begin{pmatrix} 0{\cdot}58+0{\cdot}08 & 0{\cdot}27+0{\cdot}22 \\ 0{\cdot}18+0{\cdot}16 & 0{\cdot}08+0{\cdot}46 \end{pmatrix} \begin{pmatrix} 1 & 0 \\ 0 & 1 \end{pmatrix}$$

$$= \begin{pmatrix} 0{\cdot}66 & 0{\cdot}49 \\ 0{\cdot}34 & 0{\cdot}54 \end{pmatrix} \begin{pmatrix} 1 & 0 \\ 0 & 1 \end{pmatrix}$$

$$= \begin{pmatrix} 0{\cdot}66 & 0{\cdot}49 \\ 0{\cdot}34 & 0{\cdot}54 \end{pmatrix}$$

The probability of a wet day two days after a wet day is thus 0·66, that of a wet day two days after a dry day is 0·49 and so on.

7.4 Frequency distributions

Markov chains describe the time distribution of probabilities. In the above example there was a clear bias towards a persistence of wet days, indicated by the probability $p_1 = 0{\cdot}76$. The chain was composed of a sequence of two-way probabilities. We have a simple *frequency distribution* where 76% of wet days were followed by another wet day and only 24% were followed by a dry day. This frequency distribution has been defined empirically, but we can also consider the frequency distribution of *a priori* probabilities. Returning to the examples of the dice and coin, each face of a dice has a probability of occurrence with each throw of 1/6 ($0{\cdot}1\dot{6}$), and each side of a coin will have a probability of occurrence of 0·5. A simple graph can be drawn to illustrate the frequency distribution. Such a graph, normally termed a histogram, appears in figure 7.5(a). There is no bias to cause a greater frequency for any one face over the others. Consider now the situation whereby the dice is loaded such that the frequency of occasions when the six face showed uppermost was increased

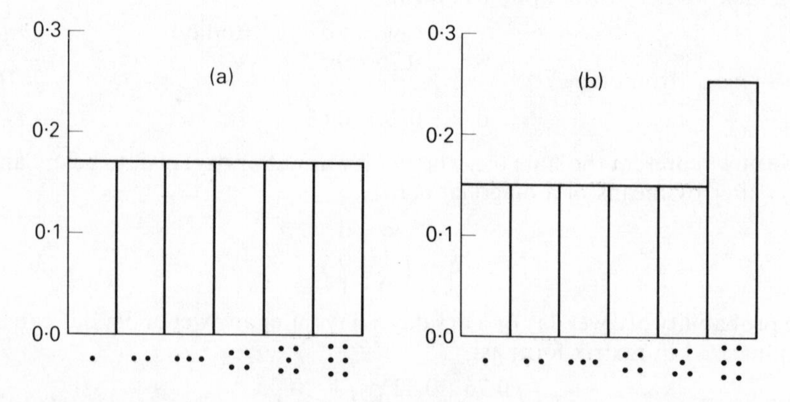

Figure 7.5 (a) Unbiased and (b) biased probabilities for a dice.

(figure 7.5(b)). The sum of all the probabilities is still 1·0, but if say $P(6)$ is found empirically to be 0·25, then the total remaining probability shared between the remaining five faces is accordingly reduced. In this case the probability of occurrence of each of the other five faces is $(1-0{\cdot}25)/5$ or 0·15.

In physical geography the frequency of occurrence of most natural events is biased. We can illustrate the bias inherent within a data set by taking a sample, normally a random sample, and plotting the frequency of events of different magnitudes on a histogram ultimately to arrive at statistical estimates of empirically derived probability. Further bias can be introduced by errors due to faulty instrumentation or inadequate sampling, which is why we require parametric statistical techniques to be able to determine the representativeness of the sample: whether or not the observed frequency distribution for the sample is likely to portray relatively accurately the frequency distribution which would result for the entire population. An example of such a histogram appears in figure 7.6. The sample in this case represents the frequency of

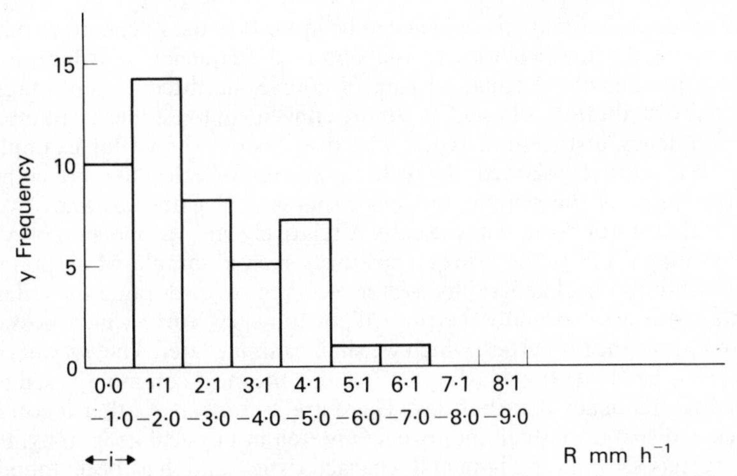

Figure 7.6 Frequency distribution of maximum hourly rainfall intensity for frontal storms at Lampeter, UK 1973–75.

maximum hourly rainfall intensities for a sequence of frontal storms in a part of the United Kingdom. Our immediate, but highly subjective judgement, since we have no way of verifying the results statistically, is that there is a clear bias towards values at the lower end of the scale. Since the overall probability must be 1·0 for the sample, the area contained within each column on the histogram is proportional to the probability of occurrence for each magnitude (0·0 to 1·0 mm h^{-1}, 1·1 to 2·0 mm h^{-1} and so on) and thus:

$$P \propto fi/N \tag{7.18}$$

where f is the frequency (number of events), i is the class interval and N is the sample size. For convenience in this particular example the class interval for rainfall intensity has been chosen as 1·0 mm h^{-1}. In many other cases however, such a convenient division is not possible, so that ultimately the total probability (ΣP) is equal to i, since:

$$\sum P = \sum \frac{fi}{N}$$

$$= \frac{i}{N} \sum f$$

But $N = \Sigma f$ and therefore:

$$\sum P = \frac{i}{N} N = i$$

It is often more useful therefore to construct the *probability density* histogram, such that:

$$P \propto fi/Ni$$

or

$$\sum P = \frac{1}{N} \sum f = 1{\cdot}0$$

The nature of the statistics which can be applied to data depends to a large extent upon the mathematics of the observed frequency distributions. In theory a mathematical equation can, of course, be fitted to any observed frequency distribution, although it is more convenient to establish a number of basic frequency distribution types. The distribution shown for example in figure 7.6 is positively skewed: that is to say the modal value does not coincide with the mean of the sample, but lies to its left. Negative skewness would indicate that the opposite was the case. A relatively large proportion of events have low magnitude in the case of a positively skewed sample. Many statistics in use within physical geography assume no skewness, adopting the criterion that the data are 'normally distributed', a non-skew curve which possesses certain important properties, which we shall examine later. The *normal curve* has already been mentioned (chapter 3) and is the most commonly used of all theoretical frequency distributions. There are a number of other theoretical frequency distributions commonly encountered in physical geography. Each of these has certain fundamental characteristics and has been found to approximate a number of observed frequency distributions. All basic parametric statistics assume the normal distribution. However, probability distributions encountered in say, river discharge studies, reveal non-normal distributions. Thus we are very much concerned with finding theoretical frequency distributions which most closely fit empirically derived histograms. The mathematical rationale behind each theoretical distribution is however strictly governed by the nature of the probability space within which the events occur and upon the assumptions made. For example, it is important to determine whether events were truly independent, and to bear in mind the sample size.

7.5 Theoretical distributions

7.5.1 The binomial distribution

Other than the normal curve this distribution is perhaps the most useful and simple of all distributions. For illustration we can return to the probability space shown in figure 7.1, where we were only concerned with the mutually exclusive, time-independent situation, implying that $P(A)+P(\sim A) = 1{\cdot}0$. For simplicity we shall continue using the example of spinning a coin and assume that there is no in-built bias, so that $p = q = 0{\cdot}5$, where p is the probability of obtaining heads, and q that of obtaining tails. The binomial distribution applies when a number of successive independent spins are made. Clearly, although with each spin the probabilities of p and q are the same, the total number of different *combinations* of faces (say two heads and one tail out of three throws) will increase as the sequence of spins increases. An *a priori* probability can be ascribed to each combination. Similarly *a priori* probabilities can be ascribed to the different *permutations* (implying order: head, head, tail or tail, tail, tail). For example, for a sequence of three throws there will be 2^3 permutations

hhh hht htt hth thh tth tht ttt

where h = head and t = tail, and therefore the probability of each permutation will be $1/2^3$ or 0·125, so that the probability of one permutation is given by $1/2^n$ where there are two alternative outcomes to each trial. More often though we are concerned with combinations of events, say the chances of two fine days out of seven, given the probability of 'fine' and 'not fine'. In the above sequence of permutations, the combination 2h + t occurs three times—hht, hth, thh—so that the probability of obtaining this combination is three times the probability for each permutation. The same is true for 2t + h, whilst there is only one way of obtaining 3h or 3t. The sequence of probabilities for the four combinations is thus, p, $3p$, $3p$, p, where p is the probability for each individual permutation.

We have so far been working 'long-hand'. Clearly what is emerging is that if the probability of obtaining one head from one spin is 0·5, then the probability of obtaining two heads from two spins is $0{\cdot}5^2$ and that of obtaining three heads from three spins $0{\cdot}5^3$ and so on. This should already have been apparent from our treatment of Markov chain probabilities. Similarly the probability of obtaining the combination 2h + t, for $p = 0{\cdot}5$, must be $0{\cdot}5^2 \times 0{\cdot}5$, or p^2q. As long as there is no bias in the probabilities of p and q the picture is simple. If however, $p = 0{\cdot}4$ and $q = 0{\cdot}6$, then we must rigorously apply the general algebraic case. This is set out below:

$$p^3+ppq+pqq+pqp+qpp+qqp+qpq+q^3 = 1{\cdot}0$$

or

$$p^3+3p^2q+3pq^2+q^3 = 1{\cdot}0$$

or

$$(p+q)^3 = 1{\cdot}0$$

In other words the sequence of probabilities is represented by the binomial expansion of $(p+q)^n$. We can thus use Pascall's triangle (chapter 6) or equations (6.13) and (6.14) to determine the probability of (say) m heads from a sequence of n spins as:

$$^{n}P_{m} = \frac{n!}{m!\,(n\text{-}m)!}p^m(1-p)^{n-m} \tag{7.19}$$

It is rare within physical geography for mutually exclusive events to have equal probabilities. More commonly we have a bias one way or the other, such that for example, the probability of obtaining bankful discharge in a river, a relatively rare occurrence, is significantly less than that for lower flows. This introduces marked positive skewness to the distribution. The frequency distribution for the binomial when $p = q = 0{\cdot}5$ is symmetrical about the mean value (figure 7.7), whilst otherwise marked skewness results. The second distribution in figure 7.7 is for $p = 0{\cdot}2$ and $(1-p) = 0{\cdot}8$. Both curves are shown for a sequence of ten trials ($n = 10$). We saw in the example developed to illustrate Markov chains that the probability (P) of there being rain on any one day was 0·57. There is thus a slight bias in this essentially binomial situation. The probability of every day in a week being wet is thus $0{\cdot}57^7 = 0{\cdot}020$—a surprisingly low probability considering the data in question are for one of the wetter parts of west Wales! The probability of there being exactly two wet days out of any seven can be calculated using equation (7.19) as follows:

$$
\begin{aligned}
{}^7P_2 &= \frac{7!}{2!(7-2)!} 0{\cdot}57^2(1-0{\cdot}57)^{7-2} \\
&= \frac{5040}{240} 0{\cdot}325 \times 0{\cdot}015 \\
&= 0{\cdot}10
\end{aligned}
$$

Note that this does *not* imply that for example, the probability of a weekend being wet is 10%, since the probability of both Saturday and Sunday being wet

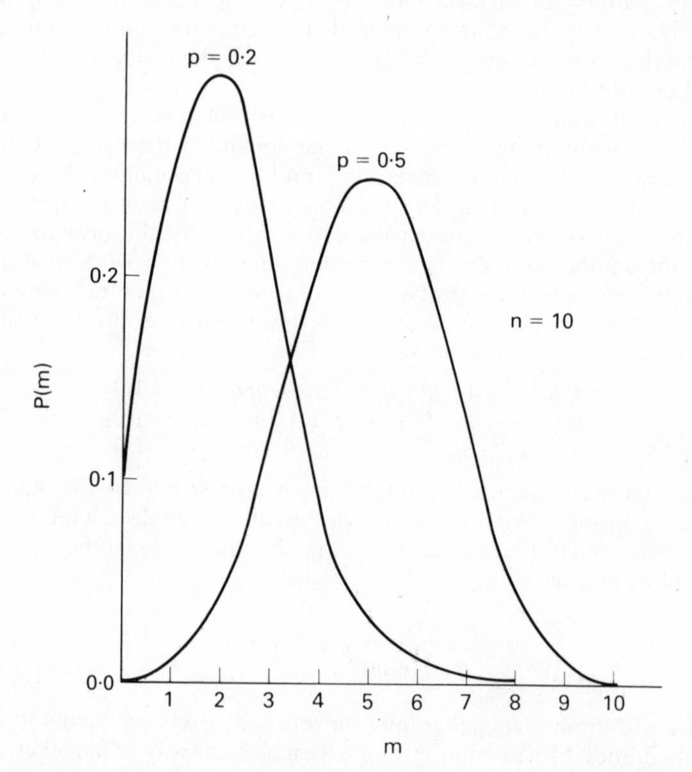

Figure 7.7 The binomial frequency distribution for $n = 10$, $p = 0{\cdot}2$ and $0{\cdot}5$.

is the permutation 0·33 ($0·57^2$). Nor does the calculation of either of these probabilities take into account the weather of previous days—there is quite a marked persistence element in the day to day weather. Taken independently the probability of *either* day of a weekend being wet is the probability ${}^2P_1$, or 0·49. In addition the probability of *at least* two days being wet out of seven is given by the sum of probabilities from ${}^7P_2$ to ${}^7P_7$, or, 0·10+0·20+0·31+0·24+0·09+0·02 = 0·96.

7.5.2 The Poisson distribution

This distribution is the limit of the binomial when the probability of an event (P) is very small, whilst the total number of trials (the sample size) is very large. It is thus useful in simulating histograms of rare events, such as floods of great magnitude, over a long period of time. Because P is very small it is a markedly skewed version of the binomial. The distribution is given by:

$$P(m) = \frac{\mu^m e^{-\mu}}{m!} \tag{7.20}$$

where $P(m)$ is the probability of m events occurring in a sequence, and μ is the mean of the frequency distribution. The sum of the probabilities of all integer values of m is 1·0. Two Poisson distributions have been plotted in figure 7.8, again using a curve to represent them; one for $\mu = 1$ and the other for $\mu = 2$. The calculated probabilities are given in table 7.1. Note that as the value of the mean increases the point at which the maximum probability occurs moves towards successively higher values of m. For higher values of μ the distribution ceases to approximate the binomial.

The Poisson distribution is useful in its application to events in time or occurrences in space which are rare and which are independent of each other. The distribution will thus provide a crude model for the form of frequency distributions produced by extracting very occasional floods from a very long data record, although the independence of such events cannot always be safely assumed. For example, if we are looking at daily discharge maxima it is quite probable that the very highest discharges occur on two or three consecutive days as a result of one major rainstorm. The scatter of rare trace elements within a soil sample might be more realistically approximated with this distribution. Suppose we were to count the number of such grains in a series of

Table 7.1 *Computed Poisson probabilities for $\mu = 1$ and $\mu = 2$.*

$\mu = 1$		$\mu = 2$	
m	$P(m)$	m	$P(m)$
0	0·37	0	0·14
1	0·37	1	0·27
2	0·18	2	0·27
3	0·06	3	0·18
4	0·02	4	0·09
5	0·003	5	0·04
6	0·0005	6	0·01

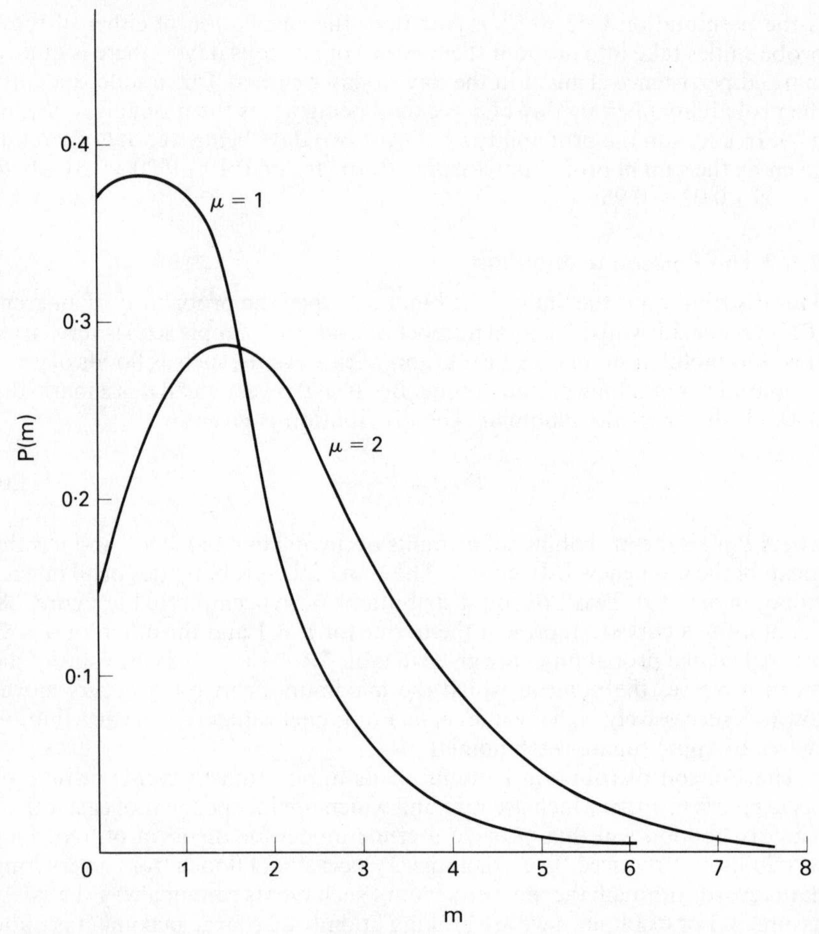

Figure 7.8 The Poisson frequency distribution for $\mu = 1$ and 2.

samples and found that there were an average 2·1 grains of a certain trace element per sample, then, using equation (7.20) we might estimate that the probability of finding at least one grain in any sample is:

$$P(1) = \frac{(2{\cdot}1)^1 e^{-2{\cdot}1}}{1!} = 0{\cdot}26$$

whilst

$$P(4) = \frac{(2{\cdot}1)^4 e^{-2{\cdot}1}}{4!} = 0{\cdot}10$$

and

$$P(10) = \frac{(2{\cdot}1)^{10} e^{-2{\cdot}1}}{10!} = 5{\cdot}6 \times 10^{-5}$$

7.5.3 The normal distribution

Just as one limit of the binomial distribution when either p or q was very small, was approximated by the Poisson distribution, so another limit of the binomial approximates the normal curve. It was noted earlier when we were dealing with the binomial distribution, that when p and q were notably different in magnitude a pronounced skewness was introduced into the distribution curve. Figure 7.7 showed a markedly skewed distribution for $p = 0{\cdot}2$ and $n = 10$, whilst a symmetrical form was produced when $p = q = 0{\cdot}5$, regardless of the number of trials. Here an important point concerning the limit to the binomial distribution occurs. Basically when p and q are equal or very nearly equal then a very few trials will produce a distribution which is very nearly symmetrical. As the difference between p and q increases we can still bring about an approximate symmetry to the distribution, but we require more and more trials. Thus, whilst the binomial $(0{\cdot}2+0{\cdot}8)^{10}$ produces a noticeably skew form, that of $(0{\cdot}2+0{\cdot}8)^{100}$ produces one nearly symmetrical about the mean. The problem is that such large expansions become unwieldy.

The normal distribution is a symmetrical curve of similar form to the limiting binomial as n increases to infinity. It has certain properties which render it one of the more useful distributions, particularly to statistics. Let us take a simple example. Imagine that simultaneous temperature measurement is made of water in a beaker using a number of thermometers in the same controlled environment. Due to minor errors in construction and design the temperature readings from each may differ slightly one to another, but the mean of all readings ($\bar{T}$) will closely approximate the actual temperature (T). The scatter of readings about the mean value will produce a distribution curve which should approximate the normal curve, with the mean at its mid-point. It will be a symmetrical curve about the mean. Any bias consistent to all thermometers will be revealed in a skewness in the distribution. This curve has characteristics which are independent of the size of sample. As with all other frequency distributions we cover in this book the area under the curve represents the total probability 1·0, and the areas under the curve between different points are proportional to the probability of occurrence of these points. With this normal curve however, the proportion of area under different parts of the curve is constant for all sample sizes, and a statistical measure (see appendix 5, section 1) is a measure of this. The probability is given in terms of the standard deviation (σ) and is:

$$P(m) = \frac{1}{\sigma\sqrt{2\pi}} \exp\left(-\frac{(m-\mu)^2}{2\sigma^2}\right) \tag{7.21}$$

Examples of the normal distribution curves for $\mu = 10$ and $\sigma = 0{\cdot}5$, 1, 2 and 5 are shown in figure 7.9. In all these the curve extends to meet the horizontal axis asymptotically, although for small standard deviations, a large proportion of cases lie very close to the mean value. Thus, as the total area under every curve must be 1·0, the smaller the standard deviation, the more peaked is the crest of the curve, with a higher percentage of points at or near the mean value. The property of the curve with respect to the standard deviation is such that there is a probability 0·682 that any item chosen at random from a normally distributed set of data will lie between $-\sigma < \mu < +\sigma$, a probability 0·954 that it will lie between $-2\sigma < \mu < +2\sigma$, and a probability 0·998 that it will lie between $-3\sigma < \mu < +3\sigma$. For most practical purposes therefore we can regard probability of occurrence beyond $\pm 3\sigma$ as negligible.

The normal distribution curve may be used to represent a very large number of naturally occurring distributions. Because of its relation to the standard

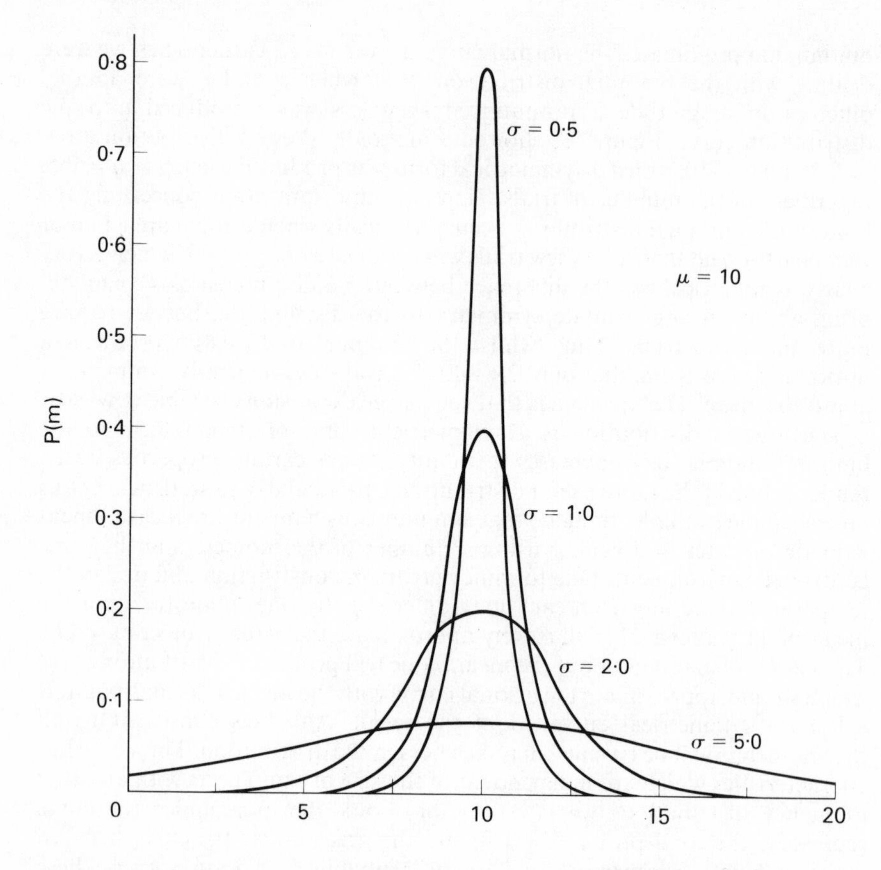

Figure 7.9 The normal frequency distribution for $\mu = 10$ and $\sigma = 0{\cdot}5$, 1, 2 and 5.

deviation, a very useful measure of the variance of data about their mean, the distribution is of primary relevance in parametric statistics, and is assumed for statistical techniques such as the product-moment correlation coefficient and least-squares regression. Any marked deviation away from the normal distribution for data to which such techniques are applied renders statistical inference unreliable. The use of the distribution as suggested above, in testing for errors, is also a major one, and is developed in the companion text (Davidson 1978). In either of these cases any marked skewness of an observed distribution from the normal may indicate that there is some systematic exterior influence upon the data. In the case of error testing this may, as indicated before, signify a consistent instrumental error. Where statistical tests which assume a normal distribution are used and the data frequency curve is skewed then certain data functions may be used to 'normalize' the curve (for example, the logarithm, or the square of individual data elements).

7.5.4 The gamma distribution

An additional very useful distribution, of particular application to data whose values are always positive, is the gamma distribution. The expression for the probability density is defined in terms of two constants, a and b, the mathematical constant e and the gamma function $\Gamma(a+1) = a!$ The complete expression for the probability is given by:

$$P(m) = \frac{m^a e^{-m/b}}{b^{a+1}\Gamma(a+1)} \tag{7.22}$$

where $b > 0$ and $a \geq -1$, and is an integer value. The probability $P(0)$ is itself zero, thus providing a clear distinction between gamma and Poisson distributions (figure 7.8). The mean of the distribution and its standard deviation can be expressed in terms of the two constants, so that for any empirically derived set of data whose mean and deviation have been calculated, the equivalent gamma distribution can be constructed for comparison, since:

$$\text{mean, } \mu = b(a+1) \tag{7.23}$$

and

$$\text{variance, } v = b^2(a+1) \tag{7.24}$$

Two examples of the gamma distribution are shown in figures 7.10(a) and (b), for $\mu = 2$ and 6 and for different variances. The curves are all asymmetrical in form, but it will be observed that those with the most marked skewness in each family of curves have the greatest variance. For a small variance the curve becomes more nearly symmetrical about the mean, as is seen in the figures when $\mu = 2$ and $v = 0{\cdot}2$. This distribution often approximates observed distributions where the vast majority of occurrences are small but not zero, for example the frequency distribution of short-period rainfall intensities within a longer rain period.

As an illustration let us calculate the gamma distribution corresponding to, say, the particle size distribution within a large sample in which there is a large range of particle sizes. Clearly there will be none of zero diameter, and experience tells us that there will be very few of large diameters. Suppose we calculate a mean particle size for our sample as $\mu = 1{\cdot}2$ mm and a standard deviation of 2·8 mm. Then, from equation (7.23) we have:

$b(a+1) = 1{\cdot}2$

and $b^2(a+1) = 7{\cdot}84$ (variance is the square of standard deviation)

so that our gamma distribution corresponding to these values is given by $a = -1$ and $b = 2{\cdot}8$. a in fact has the value $-0{\cdot}82$ from the above computation, but we must approximate it to the nearest integer value. Substituting these into equation (7.22) we obtain,

$$P(m) = \frac{m^{-1}e^{-m/2\cdot 8}}{2\cdot 8} = m^{-1}e^{-0\cdot 36m}$$

so that gamma probabilities of particles of respectively 1, 2 and 10 mm diameter are:

$$P(1) = e^{-0\cdot 36} = 0{\cdot}70$$
$$P(2) = 2^{-1}e^{-0\cdot 72} = 0{\cdot}25$$
$$P(10) = 10^{-1}e^{-3\cdot 6} = 0{\cdot}003$$

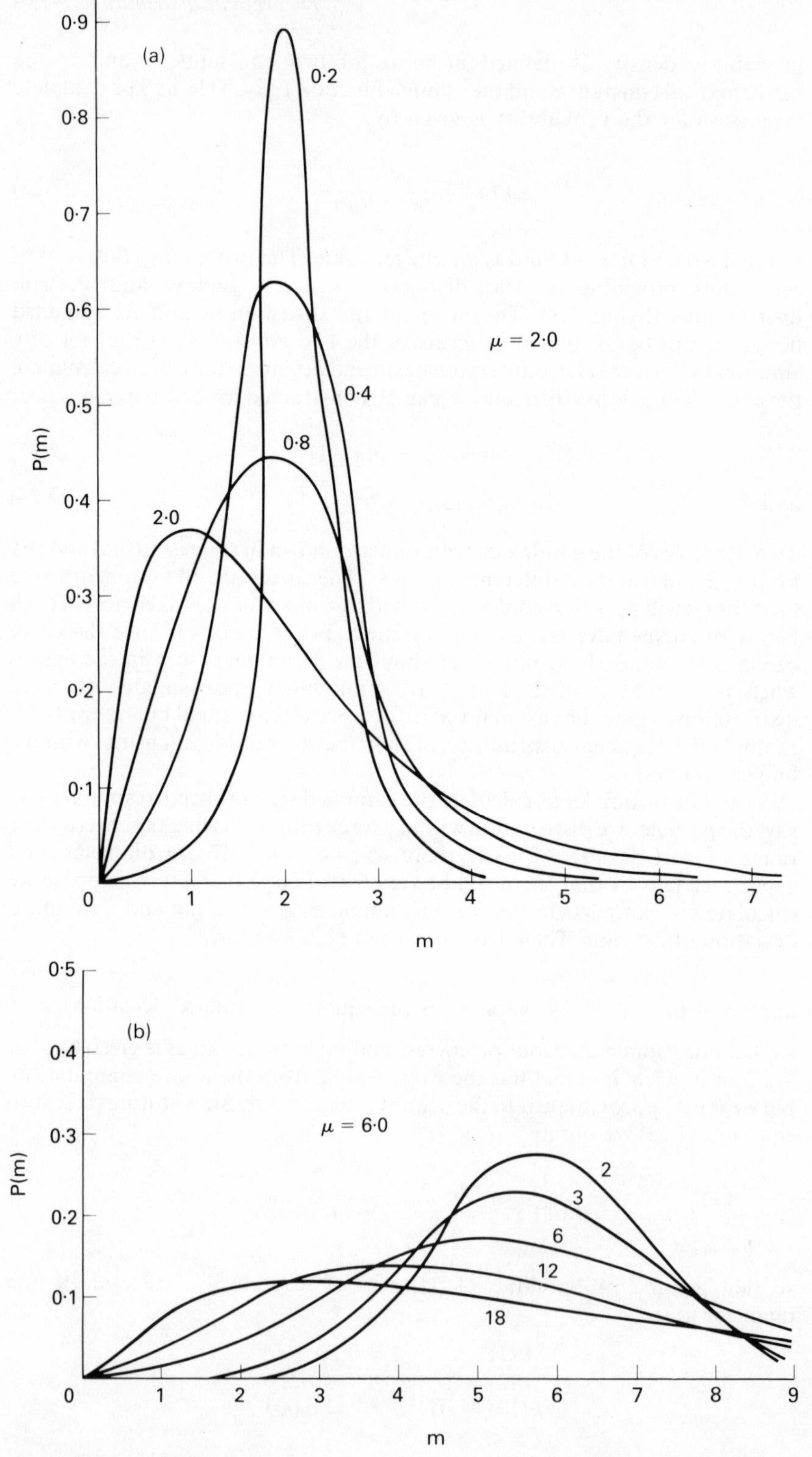

Figure 7.10 The gamma frequency distribution: (a) for $\mu = 2{\cdot}0$ and variances of 0·2, 0·4, 0·8 and 2·0; and (b) for $\mu = 6{\cdot}0$ and variances of 2, 3, 6, 12, 18.

7.6 Further types of distribution

In addition to the often-used and very useful binomial, Poisson, gamma and normal distributions, there also exist three other major groups of distributions. The derivation of these combines statistical theory with mathematics and the equations describing the probability densities are often long and appear complicated.

The first of these groups is the family of curves defined by Pearson in 1901. The Pearson family of curves comprises a system of twelve basic types of distribution, numbered using Roman numerals I to XII. Three of these are basic types between which the remaining nine are 'transitional'. The application of specific types of Pearson distributions lies largely in the field of flood hydrology. The form of all three basic types (I, IV and VI) is essentially bell-shaped but the overall skewness can be altered by changing the values of certain constants within the probability functions. For certain derived values of these constants we are able to define for example, the normal curve using one of the Pearson formulae. The scope of distributions offered by the Pearson system is very wide and correspondingly, so is the statistical rationale and mathematics behind the system. Readers interested in following up the topic should consult Elderton and Johnson (1969) as an example of a relatively straightforward mathematical and statistical explanation of the system, and Dawson (1972) for a much simplified approach involving the computer calculation of the distributions with specific reference to geography. However, for the sake of completeness a brief summary of the probability functions is given here.

7.6.1 Pearson Type I

$$P(m) = y_0\left(1+\frac{m}{a_1}\right)^{b_1}\left(1-\frac{m}{a_2}\right)^{b_2} \tag{7.25}$$

where $-a_1 < m < +a_2$, $a_1 b_2 = a_2 b_1$ and y_0 is a constant. This produces a bell-shaped curve with a marked positive skewness, unless $b_1 = b_2$ when the curve is symmetrical about the sample mean. If b_1 is negative and b_2 positive it adopts a J-shape. Davidson (1973) has used this curve to represent an observed frequency distribution of particle sizes.

7.6.2 Pearson Type IV

$$P(m) = y_0\left(1+\frac{m^2}{a^2}\right)^{-b} \exp[-\gamma \tan^{-1}(m/a)] \tag{7.26}$$

where y_0, a, b and γ are derived from two 'roots' β_1 and β_2 (see subsection 7.6.4). The curve is again characteristically bell-shaped and is normally skewed.

7.6.3 Pearson Type VI

$$P(m) = y_0(m-a)^{q_2} m^{-q_1} \tag{7.27}$$

where again y_0, a, q_1 and q_2 are defined in terms of β_1 and β_2. $q_1 > q_2$ and the curve is bell-shaped but may adopt a J-shape if q_2 is negative.

7.6.4 Other Pearson criteria and transitional types

Although there is not the space in this chapter to launch into a detailed exposition of the Pearson system all the above Pearson types and the transitional variants may be simply defined in terms of $\mathscr{K}$, where:

$$\mathscr{K} = \frac{\beta_1(\beta_2+3)^2}{4(4\beta_2-3\beta_1)(2\beta_2-3\beta_1-6)} \tag{7.28}$$

The constants β_1 and β_2, referred to in subsections 7.6.2 and 7.6.3 are defined by ratios of moments about the mean point of the distribution. The derivation of the moments about the mode is shown in figure 7.11, where the nth moment of the point m_1 about the mode is:

$$\mu_n = (m_1 - M)^n P(m_1) \tag{7.29}$$

where M is the mode, and where β_1 and β_2 are:

$$\beta_1 = \mu_3^2/\mu_2^3 \quad \text{and} \quad \beta_2 = \mu_4/\mu_2^2 \tag{7.30}$$

For the above basic distribution types we have corresponding ranges of $\mathscr{K}$:

Type I: $\mathscr{K}$ is negative
Type IV: $0 < \mathscr{K} < 1$
Type VII: $\mathscr{K} > 1$

The transitional types of curve can now be defined, again in terms of $\mathscr{K}$, and a summary of these is shown in table 7.2. Where $\mathscr{K} = 0$ and if $\beta_2 = 3$ and $\beta_1 = 0$ we have the normal curve at the transition between types I and IV. At this same point in the table, if $\beta_2 < 3$ we have the type II curve, and if $\beta_2 > 3$, the type VII curve. Of all the remaining transitional types the Pearson type III is of greatest significance to physical geography; it is a general series of curves

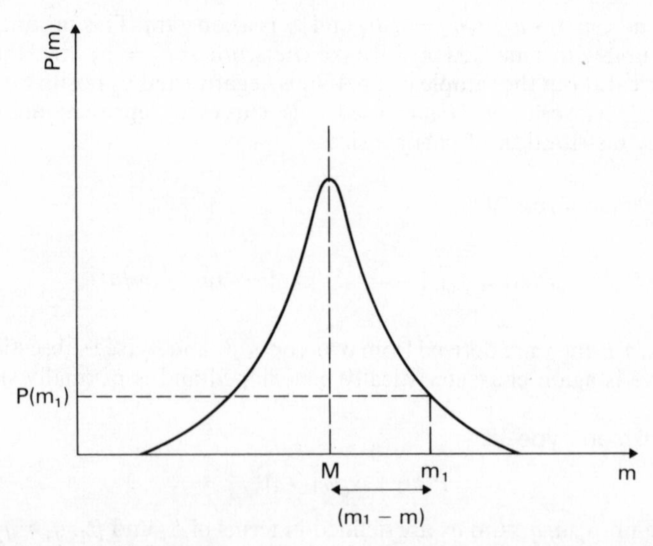

Figure 7.11 Determination of the moment of m_1 about the mode M.

Table 7.2 Pearson distributions in terms of $\mathscr{K}$. From Elderton and Johnson (1969).

$\mathscr{K} = -\infty$		$\mathscr{K} = 0$		$\mathscr{K} = 1$		$\mathscr{K} = \infty$
	Type I		Type IV		Type VI	
Type III		Type II Type VII		Type V		Type III

embracing the gamma distribution as a special case. The probability density function for the Pearson type III distribution is:

$$P(m) = p_0\left(1+\frac{m}{a}\right)^c e^{cm/a} \tag{7.31}$$

where

$$c = 4/\beta_1 - 1$$

$$a = c\mu_3/2\mu_2$$

$$P_0 = \frac{N}{a}\,\frac{c^{c+1}}{e^c\Gamma(c+1)}$$

An immediate similarity may be observed between the Pearson type III function and the gamma distribution in equation (7.22).

7.6.5 Lognormal distributions

A second important group of distributions is embraced by the *lognormal* probability. Many of the distributions previously defined are, or can be characterized by, a marked positive skewness. Throughout much of physical geography data adopt this form of distribution. However, as we saw in chapters 3 and 5, it is possible to cause a positively skewed distribution to adopt a histogram closely approximating a normal curve by taking the logarithm of the data. This is the type of transformation we apply automatically when plotting data points on logarithmic graph paper. The probability density function of this distribution is thus very similar to that of the normal distribution (equation (7.21)), and is:

$$P(m) = \frac{1}{\sigma e^{\ln x}\sqrt{2\pi}}\exp\left(-\frac{(m-\mu)^2}{2\sigma^2}\right) \tag{7.32}$$

This distribution was first used by Galton in the late nineteenth century and has since earned the name *Galton's law*. A detailed analysis of the nature of this distribution, although with special reference to economics, is to be found in Aitchison and Brown (1957).

7.6.6 Extremal distributions

The lognormal distribution above is a special case of one of three types of *extremal* or *extreme-value* distributions. In certain aspects of hydrology, fluvial geomorphology and climatology we are dealing with events, say severe floods or droughts, of extreme rareness. Frequently hydrologists for example, use the frequency of events in the past to predict similar or worse events in the future. Where we are concerned with this sort of prediction we are usually committed to utilizing *cumulative probabilities.* We shall develop this topic in more detail in section 7.7.

There are three main types of extreme value cumulative probabilities, the equations for which are given below. Firstly, there is the extremal *type I* or *Gumbel* distribution, named after a major worker in the field of extreme-value statistics. Further information on this distribution is detailed in Gumbel (1958). For this distribution:

$$P(X \leq x) = \exp(-\mathrm{e}^{-(a+x)/c}) \tag{7.33}$$

where $P(X \leq x)$ is the probability of occurrence of an event of magnitude less than x, and a and c are constants. The cumulative lognormal distribution is a special case of this distribution. The second, type II, distribution is:

$$P(X \leq x) = \exp[-(\theta/x)^k] \tag{7.34}$$

where θ is the expected highest magnitude event from a sample size n and k is a constant. Finally there is the extremal type III or *Weibull* cumulative distribution, named again after the worker who first used it. Gumbel (1958) has again developed this distribution. The cumulative probability is given by:

$$P(X \leq x) = \exp(-t^k) \tag{7.35}$$

Here

$$t = (x - \varepsilon)/(\theta - \varepsilon) \tag{7.36}$$

where k and θ are constants defined as for equation (7.34) and $-\infty < x < \varepsilon$.

7.7 The application of cumulative frequency distributions

The distributions outlined in equations (7.33) to (7.36) may appear a daunting prospect to the reader! However, the application of these may be made easier if cumulative frequency data are plotted on a relevant type of probability graph paper. We noted in chapter 5 that it is frequently easier to work with standard functions in graph form if some means of straightening the curve can be found simply by distorting the graph axes. This can also be carried out for the plots of cumulative frequency distributions. The Pearson type III, lognormal, Gumbel and Weibull distribution functions are most commonly used by the physical geographer and graph papers are printed to render such cumulative distributions into straight lines when plotted on them. An example of one of these types, the lognormal, appears in figure 7.12, and the only remaining problem facing the reader is the plotting of observational data on paper of this and similar types.

Generally, in hydrology, the *annual flood series* (or its equivalent in other branches of physical geography) is used. This series consists of the ranked annual maximum floods for the river in question, normally based upon the water year (October to September in the United Kingdom). We are concerned in this case, with the estimation of the probabilities of floods of similar or greater magnitude in future years. Put very crudely, if we have n years of data for our river, and in those n years a flood of magnitude Q has been attained m times then we may state that in the future a flood of magnitude Q or greater will

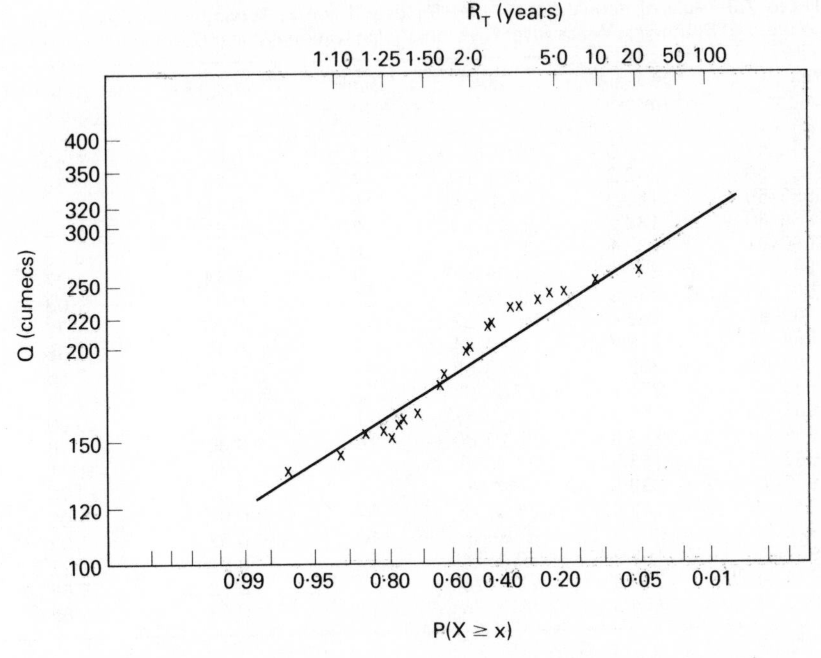

Figure 7.12 Lognormal cumulative probability plot of the data in table 7.3.

recur on average once in $(n+1)/m$ years: the *return period* or *recurrence interval.* The probability of the flood being equalled or exceeded in any one succeeding year is given by:

$$P(m) = m/(n+1) \tag{7.37}$$

Thus by ranking each annual maximum flood in descending order to give each a rank value of m we may plot Q_m against $P(m)$ on the relevant probability paper, where Q_m is discharge of rank m. This has been carried out using data from the river Tawe in south Wales (table 7.3) and the graph of the cumulative probability appears in figure 7.12 on lognormal probability paper. We could carry out a similar exercise on, for example, extremely low discharges in a similar manner, except that the ranking would be in *ascending* order, since we should be concerned with a discharge of a certain magnitude *not* being exceeded.

In either case a probability of 0·05 or 5% represents a return period of once every 20 years, indicating that for the flood series, the event of magnitude Q will be equalled or exceeded on average once every 20 years. Of course it is quite possible for such an event to occur on every one of five consecutive years, and then not to occur again for another 100. The probability of an event of magnitude Q, rank m, occurring once in a sequence of N years can be derived

Table 7.3 Table of annual flood series for river Tawe at Ystradgynlais, Wales. Data by courtesy of Southwest Wales River Division, Welsh National Water Development Authority.

Year	Discharge (m^3s^{-1})	Date	Rank m	$P=m/(n+1)$	Return period (years)
1956–57	179·5	6-9-57	13	0·63	1·59
1957–58	242·2	23-9-58	4	0·22	4·55
1958–59	156·9	17-1-59	16	0·77	1·30
1959–60	186·2	25-11-59	12	0·61	1·64
1960–61	233·1	3-12-60	6	0·32	3·13
1961–62	217·5	11-9-62	9	0·44	2·27
1962–63	134·9	9-3-63	20	0·97	1·03
1963–64	159·4	19-11-63	15	0·76	1·32
1964–65	219·7	13-12-64	8	0·42	2·38
1965–66	233·1	18-12-65	7	0·35	2·86
1966–67	251·5	27-2-67	2	0·11	9·09
1967–68	255·3	19-10-67	1	0·05	20·00
1968–69	153·9	28-10-68	18	0·86	1·16
1969–70	163·5	21-1-70	14	0·71	1·41
1970–71	236·8	1-11-70	5	0·27	3·70
1971–72	142·4	15-2-72	19	0·92	1·09
1972–73	202·4	5-8-73	10	0·52	1·92
1973–74	153·9	30-1-74	17	0·83	1·21
1974–75	243·9	22-1-75	3	0·18	5·56
1975–76	198·3	1-12-75	11	0·53	1·89

using the general binomial expansion for $P(m)$ and $(1-P(m))$ (equation(7.19)):

$$\text{Probability} = \frac{N!}{(N-1)!}P(m)^1(1-P(m))^{N-1}$$

$$= NP(m)(1-P(m))^{N-1} \tag{7.38}$$

The probability of an event occurring in exactly N years time is similarly given by:

$$\text{Probability} = P(m)(1-P(m))^{N-1} \tag{7.39}$$

So, if we were to estimate the probability of the discharge of magnitude Q occurring over a ten-year period, then this would be given by: $10 \times 0{\cdot}04 \times 0{\cdot}96^9$ for $P(m) = 0{\cdot}04$ (recurrence interval once in 25 years). There is thus a probability of 0·28 that the '25-year flood' will occur in any period of ten consecutive years. The results obtained will differ slightly according to the type of cumulative distribution selected as the model, although probably not significantly unless the plot of ranked points on the graph differs very much from that of the straight line. There is not the space to illustrate these differences here and the reader is referred to Linsley, Kohler and Paulus (1975) for an example comparing the Log–Pearson and Gumbel, and to Chow (1964) for a more detailed treatment of cumulative distributions within hydrology.

8

Integration

The contents of this chapter are natural developments of the processes of differentiation outlined in chapter 5, and use is made of expansions of series as defined in chapter 6.

8.1 Limits and continuous functions

We have seen that certain geometric series converge to limits when they are of the form $a, ar^2, ar^3, \ldots$ and $-1 < r < +1$. The sum of such a series is finite for an infinite number of terms. We can therefore write:

$$\lim_{m \to \infty} S_m = \frac{a}{1-r}$$

the limit of the expansion of S_m as the number of terms approaches infinity. Similarly in chapter 5 we cited some examples of mathematical functions of x where $f(x)$ approached a finite value as the value of x approached infinity. Consider the function $y = 1/x^2$. As the value of x is increased in integer steps then y decreases to ever smaller magnitude, approaching but never actually reaching zero:

$x =$	1	2	3	4	5	6	7	...
$y =$	1	1/4	1/9	1/16	1/25	1/36	1/49	...

We can write this in a form similar to that for the series sum:

$$\lim_{x \to \infty} y = 0$$

or the limit of y as x tends to infinity is zero. For values of x where $0 < x < \infty$ we are able to calculate values of y. The function $y = 1/x^2$ is thus said to be *continuous* between these limits.

Now let us consider a function of the form:

$$y = (x^2 - 1)/(x - 1)$$

We can compute the corresponding values of y for integer increements in x as we did in the previous example of $y = 1/x^2$. Taking again the sequence from $x = 0$ to $x = 7$ we have the following values for y:

$x =$	0	1	2	3	4	5	6	7
$y =$	1	0	3	4	5	6	7	8

The value of y when $x = 1$ is shown as zero; incorrectly as we shall see later. We gain the impression of a linear relationship apart from this interruption to the sequence. However, on closer inspection we can see that the function is not a simple linear one. The table below shows calculated values for y when x values are close to 1·0:

$x =$	0·5	0·8	0·9	0·95	0·99	0·999	1	1·001	1·01	1·05	1·1	1·2	1·5
$y =$	1·5	1·8	1·9	1·95	1·99	1·999	?	2·001	2·01	2·05	2·1	2·2	2·5

Clearly the value of y converges from both sides towards the value 2·0 but never actually reaches it, so here we must say:

$$\lim_{x \to 1} y = 2{\cdot}0$$

or

$$\lim_{x \to 1} f(x) = 2{\cdot}0$$

The function again has limits but is also *discontinuous.* If we substitute the value of 1 for x we arrive at the solution for y of 0/0, which is in fact a meaningless term, and we would be incorrect to give y the value zero, as we would if we were to give it any other value. When we are dealing with both differentiation and integration, the subject of this chapter, it is important to stress that functions should be continuous: that is, valid for all ranges of x and y within given limits.

8.2 Antiderivatives

We are deliberately terming the heading for this section 'antiderivatives' since these are precisely what integrals are! We dealt at length in chapter 5 with differentiation and derivatives and must now add that the integral of a function is another function whose derivative is the same as the original function—the antiderivative. This is the *fundamental law of calculus*, and links elementary differential calculus to the integral calculus introduced here.

We have yet to define however what the process of integration actually involves, or indeed what it may be used for and what it means. At this stage in the chapter we shall restrict our terms of reference to what are known as *definite integrals*, integrals of functions defined within known limits. Essentially the integral of a function provides us with an expression of the *area* contained beneath the graph of that function within specified bounds, and generally above the x-axis ($y = 0$).

Let us by way of introduction return to the example given in chapter 5, of the function relating infiltration rate into a soil to time, as suggested by Philips (1957–58). To avoid confusion later in the chapter we shall revert to the use of x and y rather than use t and i as we did in chapter 5. Suppose that for a given site the simple relation of this form is:

$$y = 15 + 5x^{-1/2} \tag{8.1}$$

This equation will permit us to calculate the instantaneous infiltration rate at any point in time as long as x is greater than zero, so that we have:

$x =$	0·1	0·2	0·5	1·0	2·0	5·0	h
$y =$	30·8	26·2	22·1	20·0	18·5	17·3	mm h^{-1}

The *total* quantity of water (Q) which has infiltrated the soil in any given time period is given by the area beneath the curve between the limits of the time period on the x-axis. Dimensionally this is correct as the quantity of water (in mm rainfall equivalent) thus equals mm h^{-1} h = mm. Suppose now that we wish to obtain a figure for the total quantity of water infiltrated between 0·1 and 0·5 hours. We could of course laboriously measure the area off the graph (figure 8.1) between these two points (shaded). The process of integration however exists to make this task easier, but is probably best illustrated at this stage with direct reference to geometric figures.

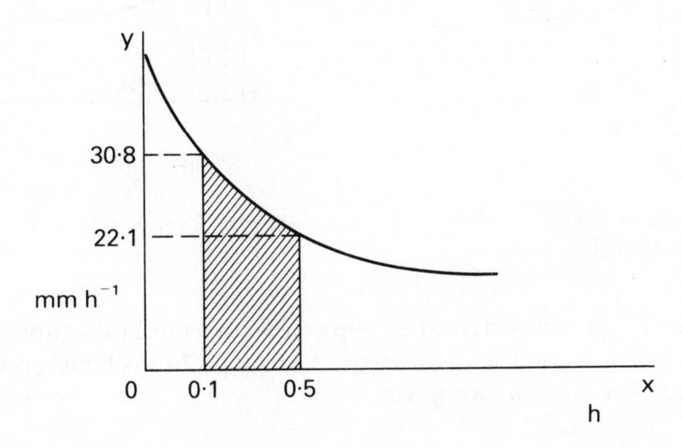

Figure 8.1 Area under the curve $y = 15 + 5x^{-\frac{1}{2}}$ from $x = 0{\cdot}1$ to $x = 0{\cdot}5$.

The area of the rectangle bounded by $x = 0{\cdot}1$, $x = 0{\cdot}5$, $y = 0$ and $y = 22{\cdot}1$ (the value of y at $x = 0{\cdot}5$) is of course easily calculated, but we are still faced with the problem of determining the area of the irregular figure between this and the curve itself. We can approach this problem by viewing the area of the rectangle as the worst possible approximation of the area under the curve between those limits. We can improve our approximation by taking the sum of a number of smaller rectangles whose upper bounds are conditioned by the points at which they contact the curve. Any number of such rectangles can be drawn in (figure 8.2(a)) between the specified limits, and the more we have the closer does our approximation approach the true value for the area under the curve between those limits. Our approximation would however be an underestimate except at the limit where the number of columns became infinitely numerous and infinitely narrow. Another way of estimating area would be to draw our columns in such a way that they provided an *overestimate* of the area, as with figure 8.2(b). Again the narrower our columns, the closer we should be to calculating the exact area. Both these approximations converge to a limit which is the true integral as δx tends to 0. We thus have:

for case (a)
$$\sum_{0\cdot1}^{0\cdot5} Q = \sum_{0\cdot1}^{0\cdot5} y\,\delta x \qquad (8.3)$$

for case (b)
$$\sum_{0\cdot1}^{0\cdot5} Q = \sum_{0\cdot1}^{0\cdot5} (y+\delta y)\delta x \qquad (8.4)$$

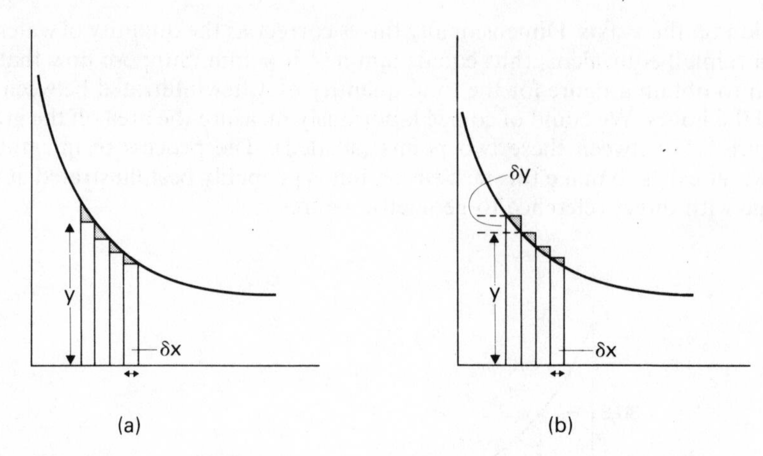

Figure 8.2 Integration of areas under a curve: (a) underestimates, equation (8.3); (b) overestimates, equation (8.4).

As an illustration of the areas arrived at using these two methods, suppose that we select the value for δx of 0·1, so that we have four columns between $x = 0{\cdot}1$ and $x = 0{\cdot}5$, then for case (a) we have:

$$\sum_{0\cdot1}^{0\cdot5} y\,\delta x = 0{\cdot}1[15+5(0{\cdot}2^{-\frac{1}{2}})]+0{\cdot}1[15+5(0{\cdot}3^{-\frac{1}{2}})]+0{\cdot}1[15+5(0{\cdot}4^{-\frac{1}{2}})]$$

$$+0{\cdot}1[15+5(0{\cdot}5^{-\frac{1}{2}})]$$

$$= 2{\cdot}62+2{\cdot}41+2{\cdot}29+2{\cdot}21 = 9{\cdot}53\text{ mm}$$

Similarly for case (b) we have:

$$\sum_{0\cdot1}^{0\cdot5} (y+\delta y)\delta x = 0{\cdot}1[15+5(0{\cdot}1^{-\frac{1}{2}})]+0{\cdot}1[15+5(0{\cdot}2^{-\frac{1}{2}})]+0{\cdot}1[15+5(0{\cdot}3^{-\frac{1}{2}})]$$

$$+0{\cdot}1[15+5(0{\cdot}4^{-\frac{1}{2}})]$$

$$= 3{\cdot}08+2{\cdot}62+2{\cdot}41+2{\cdot}29 = 10{\cdot}40\text{ mm}$$

We have deliberately chosen very large value for δx and our resulting estimates for area are thus rather crude with a discrepancy of nearly one millimetre between the two extremes. All we do know at the present stage is that the true value for the area lies between these two values.

As $\delta x \to 0$ the two estimates converge to the true value, and it is this value which is the *definite integral* of y with respect to x, between $x = 0{\cdot}1$ and $x = 0{\cdot}5$, or we can write:

$$\lim_{\delta x\to 0}\sum_{0\cdot1}^{0\cdot5} y\delta x = \int_{0\cdot1}^{0\cdot5} y\,\mathrm{d}x \tag{8.4}$$

This definite integral is the antiderivative of the function $y = 15+5x^{-1/2}$, the expression which will differentiate to produce the function. In general terms we write the integral $\int f(x)\mathrm{d}x$ for the function $y = f(x)$ and we know that:

$$y\delta x < \delta A < (y + \delta y)\delta x \tag{8.5}$$

where A is the area, and A increases by δA as x increases by δx. Thus, dividing (8.5) through by δx we have:

$$y < \frac{\delta A}{\delta x} < y + \delta y$$

and taking the limit as $\delta x \to 0$ and $\delta y \to 0$ we have:

$$\frac{\mathrm{d}A}{\mathrm{d}x} = y \tag{8.6}$$

We thus require some means of integrating the function $y = 15 + 5x^{-1/2}$ to arrive at a true value for the definite integral $\int_{0.1}^{0.5}(15 + 5x^{-1/2})\mathrm{d}x$. It is to this problem that sections 8.4 to 8.11 are devoted.

8.3 The indefinite integral

We have so far considered integrals only as finite measures, essentially of area, by applying limits and solving for their definite forms. *Indefinite integrals* are the more general cousins, having no such defined limits. For example, since if $y = 5x^2 + 3x + 2$ $\mathrm{d}y/\mathrm{d}x = 10x + 3$, then the reader may conclude that by definition we may similarly write:

$$\int (10x + 3)\mathrm{d}x = 5x^2 + 3x + 2$$

This is, on closer inspection, only one of an infinite number of antiderivatives of the function, and the integration could equally well be $5x^2 + 3x + 1$, or $5x^2 + 3x + 2{,}000$. The constant value of the antiderivative is undefined. When we considered integration between limits there was clearly no need for the addition of a constant value each time, as this would make no difference to the overall result of the integration: the constant term being removed in the operation. For example, to determine the area beneath the curve in the equation $y = 10x + 3$ between the limits of $x = 5$ and $x = 10$ we should simply write:

$$\int_5^{10} (10x + 3)\mathrm{d}x = \Big[5x^2 + 3x + C\Big]_5^{10}$$
$$= (500 + 30 + C) - (125 + 15 + C)$$
$$= 390$$

Note the convention that the integral is contained within square brackets prior to the imposition of limits. Knowing the integral we can simply substitute in the limits of x and subtract one result from the other. In general therefore we must write for indefinite integrals:

$$\int f(x)\mathrm{d}x = g(x) + C \tag{8.7}$$

8.4 The standard integrals

It should by now be clear that since we are able to define a number of standard derivatives of functions (chapter 5), the corollary must be that for each derivative there is an analogous antiderivative, or *standard integral.* Thus, just as we were able to write:

$$\frac{\mathrm{d}}{\mathrm{d}x}(ax^n) = nax^{n-1}$$

we can similarly write:

$$\int ax^n\,\mathrm{d}x = \frac{ax^{n+1}}{n+1} + C \tag{8.8}$$

since:

$$\frac{\mathrm{d}}{\mathrm{d}x}\left(\frac{ax^{n+1}}{n+1}\right) = \frac{ax^{(n+1)-1}}{n+1}(n+1) = ax^n$$

Again as:

$$\frac{\mathrm{d}}{\mathrm{d}x}\sin x = \cos x$$

then:

$$\int \cos x\,\mathrm{d}x = \sin x + C$$

and so on. A list of some standard integrals appears below. A more complete list is of book length in itself and the reader should refer to Dwight (1957) for other standard integrals. Those mentioned here are sufficient for an understanding of the remainder of this chapter:

$$\int x^n\,\mathrm{d}x = \frac{x^{n+1}}{n+1} + C \quad (x \neq -1) \tag{8.9}$$

$$\int (1/x)\,\mathrm{d}x = \ln x + C \tag{8.10}$$

$$\int \mathrm{e}^x\,\mathrm{d}x = \mathrm{e}^x + C \tag{8.11}$$

$$\int \sin x\,\mathrm{d}x = -\cos x + C \tag{8.12}$$

$$\int \cos x\,\mathrm{d}x = \sin x + C \tag{8.13}$$

$$\int \sec^2 x\,\mathrm{d}x = \tan x + C \tag{8.14}$$

$$\int \operatorname{cosec}^2 x\,\mathrm{d}x = -\cot x + C \tag{8.15}$$

$$\int \sinh x\,\mathrm{d}x = \cosh x + C \tag{8.16}$$

$$\int \cosh x\,\mathrm{d}x = \sinh x + C \tag{8.17}$$

$$\int \operatorname{sech}^2 x\,\mathrm{d}x = \tanh x + C \tag{8.18}$$

$$\int 1/\sqrt{(a^2 - x^2)}\,\mathrm{d}x = \sin^{-1}(x/a) + C \tag{8.19}$$

$$\int 1/\sqrt{(a^2-x^2)}\,dx = \cos^{-1}(x/a)+C \qquad (8.20)$$

$$\int 1/(a^2+x^2)\,dx = \frac{1}{a}\tan^{-1}(x/a)+C \qquad (8.21)$$

$$\int 1/\sqrt{(a^2+x^2)}\,dx = \frac{1}{a}\sinh^{-1}(x/a)+C \qquad (8.22)$$

$$\int 1/\sqrt{(x^2-a^2)}\,dx = \frac{1}{a}\cosh^{-1}(x/a)+C \qquad (8.23)$$

$$\int a/(a^2-x^2)\,dx = \frac{1}{a}\tanh^{-1}(x/a)+C \qquad (8.24)$$

$$\int \log x\,dx = x\log x - x + C \qquad (8.25)$$

Problems 8.4

Obtain the solutions to the following integrals:

8.4.1 $\int (5x^2+3x-2)dx$

8.4.2 $5\int (3x^2-5+2x)dx$

8.4.3 $\int_{\pi}^{2\pi} (\sin x - \cos x)dx$

8.4.4 $\int_{1}^{3} (3/x+2x^2)dx$

8.4.5 $\int \frac{1}{1-x^2}dx$

8.4.6 $\int_{-5}^{2} (3x^3-2x)dx$

8.4.7 $\int_{1}^{5} \frac{dx}{x}$

8.4.8 $\int (\sec^2 x - 2\cos x)dx$

8.5 Integration of more complicated functions

On many occasions when integrals have to be found, the form of the original function is such that only rarely is it possible to integrate using the above standard forms directly. Thus a number of different methods have been derived to assist us in the task of determining integrals of a more complicated nature. Very often integrals can be deduced by a process not very far removed from trial and error. A hunch may be played by making an educated guess as to what the integral might be and this result is then differentiated (a generally easier process) for comparison with the original. However, there are three major methods for deriving the integrals of more complicated functions, which are of a more sophisticated nature. These are:

(1) integration by parts
(2) integration by substitution
(3) integration using partial fractions

All these apply to the solution of indefinite integrals as well as ultimately to definite (numerical) integration, a topic developed further later in the chapter. These three main methods are outlined in the next four sections, and are summarized in table form in section 8.10.

First of all however, we should consider the simple case of finding the integral of functions of the form:

$$\int f(Ax+B)\mathrm{d}x$$

A general rule applies to functions of this form so that:

$$\int f(Ax+B)\mathrm{d}x = \frac{1}{A}F(Ax+B)+C \tag{8.26}$$

This is very useful when we are required to integrate a function such as $\sec^2(5x+2)$, since we can simply write:

$$\int \sec^2(5x+2)\mathrm{d}x = \tfrac{1}{5}\tan(5x+2)+C \qquad \text{(from 8.14))}$$

Should it not be possible to integrate the function in question by this method then one of the more sophisticated methods outlined below must be used. The choice of method is often very much a matter of noticing that a function has a particular form. For this reason the summary of 'rules' of integration in section 8.10 should be memorized and sections 8.5 to 8.9 studied in detail, attempting all problems. If the function to be integrated is of the form $\int x \log x \,\mathrm{d}x$, composed of the product of two functions then the integral can be found by the method outlined in section 8.6, whilst if it is of the form of a polynomial, then its solution should be approached using the technique in section 8.9. In some cases integration can be approached by using more than one method, hopefully yielding the same result regardless of the technique used! Perhaps one of the most useful methods is that of integration by parts.

8.6 Integration by parts

We noted in chapter 5 that we could differentiate functions of the form $y = u(x)v(x)$ by using the formula given in equation (5.33):

$$\frac{\mathrm{d}y}{\mathrm{d}x} = u(x)\frac{\mathrm{d}y}{\mathrm{d}x} + v(x)\frac{\mathrm{d}u}{\mathrm{d}x}$$

For the sake of clarity let us simplify the form of the equation and write:

$$\frac{\mathrm{d}}{\mathrm{d}x}(uv) = u\frac{\mathrm{d}y}{\mathrm{d}x} + v\frac{\mathrm{d}u}{\mathrm{d}x} \tag{8.27}$$

By integrating the entire expression with respect to x we obtain:

$$uv = \int u\frac{dv}{dx}dx + \int v\frac{du}{dx}dx + C$$

so that therefore:

$$\int u\frac{dv}{dx}dx = uv - \int v\frac{du}{dx}dx + C \qquad (8.28)$$

or

$$\int u\,dv = uv - \int v\,du + C \qquad (8.29)$$

For example, suppose we have to derive the following integral:

$$\int x \sin x\, dx$$

Let $u = x$ and let $dv = \sin x\, dx$, so that $v = -\cos x$, and thus:

$$\int x \sin x\, dx = -x\cos x - \int -\cos x\, dx \qquad \text{(from (8.29))}$$

$$\int x \sin x\, dx = \sin x - x\cos x + C$$

The operation is thus quite easy so far. Now, however, suppose we have to integrate the function:

$$\int x^2 \sin x\, dx$$

If we let $u = x^2$ and let $dv = \sin x$, so that $du = 2x\, dx$ and $v = -\cos x$ then:

$$\int x^2 \sin x\, dx = -x^2\cos x - \int -2x\cos x\, dx$$

and we are faced with another integral which we are unable to solve immediately. We thus have to integrate this expression again with respect to x, again using integration by parts. Let us now substitute $s = 2x$ and $dt = -\cos x\, dx$, so that $ds = 2\, dx$ and $t = -\sin x$, then:

$$\begin{aligned}\int -2x\cos x\, dx &= -2x\sin x - \int -2\sin x\, dx \\ &= -2(\cos x + x\sin x) + C\end{aligned}$$

Thus the complete integral is:

$$\begin{aligned}\int x^2 \sin x\, dx &= -x^2\cos x + 2x\sin x + 2\cos x + C \\ &= \underline{2x\sin x + (2 - x^2)\cos x + C}\end{aligned}$$

Note here that if we had let $s = -\cos x$ and $dt = 2x\,dx$, so that $ds = \sin x\,dx$ and $t = x^2$ then we should have:

$$\int -2x\cos x\,dx = -x^2\cos x - \int x^2 \sin x\,dx$$

and we should have for the original integral:

$$\int x^2 \sin x\,dx = -x^2\cos x + x^2\cos x + \int x^2 \sin x\,dx$$

or $$\int x^2 \sin x\,dx = \int x^2 \sin x\,dx\ !!$$

Clearly then we must use some common sense when applying integration by parts and select which of the parts of the function is to be represented by u and which by dv. The best way to perfect the technique is to try out some integrations using this method on your own. As a start it is suggested that you should try to verify the following equalities:

$$\int x^2\cos x\,dx = x^2\sin x - 2(\sin x - x\cos x) + C$$

$$\int x^2\log x\,dx = \tfrac{1}{3}x^3(\log x - \tfrac{1}{3}) + C$$

$$\int x^2 e^x\,dx = x^2e^x - 2xe^x + 2e^x + C$$

$$\int \arcsin x\,dx = x\arcsin x + (1-x^2)^{1/2} + C$$

Now attempt the following problems.

Problems 8.6

Integrate:

8.6.1 $\int 5x\log x\,dx$

8.6.2 $\int x^3\sin x\,dx$

8.6.3 $\int \log x\,dx$

8.6.4 $\int (\log x)^2\,dx$

8.6.5 $\int 3x^2(1-\log x)dx$

8.7 Integration by substitution

Another major technique of integration, again where we are unable to use standard integrals directly, is by substitution. Here we use a property similar to that used in integration by parts, that:

$$dx = \frac{dx}{du}du \tag{8.30}$$

Thus, if we substitute u for a part of the integrand, we can solve for du/dx and then find dx/du. To illustrate the technique we shall look at some worked examples of integrals derived in this way. Firstly let us attempt to find the integral:

$$\int \frac{x}{\sqrt{(1-x)}}dx$$

Let $u = \sqrt{(1-x)}$. The derivative of this can be found by the method of substitution outlined in section 6.7. Let $t = 1-x$,

$$\frac{dt}{dx} = -1$$

Also we have $$\frac{du}{dt} = \tfrac{1}{2}t^{-1/2}$$

so that, $$\frac{du}{dx} = \frac{du}{dt}\frac{dt}{dx} = -\tfrac{1}{2}(1-x)^{-1/2}$$

$$\therefore \frac{dx}{du} = -2(1-x)^{1/2}$$

and that, $$dx = -2(1-x)^{1/2}\,du$$

So that, substituting in for du we have:

$$\int \frac{xdx}{\sqrt{(1-x)}} = \int \frac{x}{(1-x)^{\frac{1}{2}}}[-2(1-x)^{\frac{1}{2}}]du$$

and $$\int \frac{xdx}{\sqrt{(1-x)}} = \int -2x\,du$$

Now if $u = (1-x)^{\frac{1}{2}}$, $u^2 = (1-x)$ and thus $x = 1-u^2$. Substituting this into the integral we have:

$$\int -2(1-u^2)du = -2u + \tfrac{1}{3}u^3 + C$$

Resubstituting x for u we have:

$$\int \frac{x}{\sqrt{(1-x)}}dx = -2(1-x)^{\frac{1}{2}} + \tfrac{1}{3}(1-x)^{\frac{3}{2}} + C$$

$$= \underline{-\tfrac{2}{3}(1-x)^{\frac{1}{2}}(x+2) + C}$$

For a further example let us find the integral:

$$\int xe^{x^2}dx$$

Here let $u = e^{x^2}$ and let $t = x^2$, so that $u = e^t$

$$\frac{du}{dt} = e^t \quad \text{and} \quad \frac{dt}{dx} = 2x$$

$$\frac{du}{dx} = \frac{du}{dt}\frac{dt}{dx} = 2xe^{x^2}$$

So that,
$$\frac{dx}{du} = 1/2xe^{x^2}$$

Thus,
$$\int xe^{x^2}\,dx = \int xe^{x^2}(1/2xe^{x^2})du$$

$$= \int \tfrac{1}{2}\,du = \tfrac{1}{2}u + C$$

and
$$\int xe^{x^2}\,dx = \underline{\tfrac{1}{2}e^{x^2} + C}$$

As a final example, let us find:

$$\int \frac{\log x}{x}\,dx$$

Let $u = \log x$ so that,

$$\frac{du}{dx} = \frac{1}{x}$$

Thus
$$\int \frac{\log x}{x}\,dx = \int \frac{\log x}{x}\,x\,du$$

$$= \int \log x\,du$$

Now since, if $u = \log_a x$, then $x = a^u$ and so $\log_a a^u = u$, we can write:

$$\int u\,du = \int \log x\,du = \tfrac{1}{2}u^2 + C$$

and
$$\int \frac{\log x}{x}\,dx = \underline{\tfrac{1}{2}(\log x)^2 + C}$$

The method is particularly useful where one part of the integrand can be expressed in terms of the derivative of the other, as in this last example. So that in the case of finding $\int f_1(x)f_2(x)\,dx$, where $f_2(x) = f_1'(x)$, we let $u = f_1(x)$ and then solve the integrand:

$$\int f_1(x)f_2(x)\,dx = \int f_1(x)f_2(x)\frac{dx}{du}du$$

$$= \int f_1(x)f_2(x)\frac{1}{f_1'(x)}du$$

$$= \int f_1(x)du$$

and substituting u for $f_1(x)$.

Again as with the previous section, the best way of perfecting the technique is to work through some examples. The reader should verify the four integrals below and their solutions, using the method of substitution:

$$\int \cos 2x\,dx = \tfrac{1}{2}\sin 2x + C$$

$$\int \frac{x^2\,dx}{(x-1)^4} = \frac{-x^2+x-\frac{1}{3}}{(x-1)^3} + C$$

$$\int \frac{1}{x\log x}dx = \log(\log x) + C$$

$$\int \frac{e^x\,dx}{e^x+1} = \log(e^x+1) + C$$

Problems 8.7

Integrate the following by the method of substitution:

8.7.1 $\int 5\sin x\cos x\,dx$

8.7.2 $\int (1+x^2)^{-1}\tan^{-1}x\,dx$

8.7.3 $\int (1-x)^{\frac{1}{2}}x\,dx$

8.7.4 $\int \tan x\sec^2 x\,dx$

8.7.5 $\int x^2 e^{-2x^3}dx$

8.8 The integration of trigonometric functions

A further type of substitution in the process of integration can be made where trigonometric functions appear in the integrands. Some such examples have already been integrated in the previous section, though not in this way. A number of equations were given in chapter 2 (equations (2.8) to (2.20)) relating the trigonometric functions of double angles ($\sin 2x$, $\cos 2x$, etc.) to similar functions for single angles, and the squares of trigonometric ratios. These types of substitution may be used wherever appropriate in integration. Subsequently the integration can be performed by parts or by substitution or other methods. For example, if we take the integrand:

$$\int \sin^2 x\,dx$$

we know that from equation (2.20):

$$\cos 2x = \cos^2 x - \sin^2 x$$

and from (2.8) we have:

$$\sin^2 x + \cos^2 x = 1$$

and thus:

$$\sin^2 x = \tfrac{1}{2}(1 - \cos 2x)$$

Substituting this into the original integrand we have:

$$\begin{aligned}\int \sin^2 x\,dx &= \tfrac{1}{2}\int(1-\cos 2x)\,dx \\ &= \tfrac{1}{2}\int dx - \tfrac{1}{2}\int \cos 2x\,dx \\ &= \tfrac{1}{2}x - \tfrac{1}{4}\sin 2x + C\end{aligned}$$

It is assumed here that the reader will already have proved to his own satisfaction that:

$$\int \cos 2x\,dx = -\tfrac{1}{2}\sin 2x \quad \text{(from an exercise in section 8.7)}$$

In fact we may use a general form of integral for the integration of all powers of trigonometric ratios, so that where n is any positive integer we have:

$$\int \sin^n x\,dx = -\frac{1}{n}\sin^{n-1}x\cos x + \frac{n-1}{n}\int \sin^{n-2}x\,dx \tag{8.31}$$

and

$$\int \cos^n x\,dx = \frac{1}{n}\cos^{n-1}x\sin x + \frac{n-1}{n}\int \cos^{n-2}x\,dx \tag{8.32}$$

These formulae are called the *reduction formulae* and are used as many times as are necessary in order to produce a final integrand which is immediately integrable, of the form $\sin x$ or $\cos x$. As an exercise in the use of these formulae it is suggested that the reader again attempts the problems in integration below:

$$\int \sin^4 x\,dx = \tfrac{1}{4}\sin^3 x\cos x - \tfrac{5}{8}\sin x\cos x + \tfrac{3}{8}x + C$$

$$\int \cos^2 x\,dx = \tfrac{1}{2}\cos x\sin x + \tfrac{1}{2}x + C$$

$$\int \cos^5 x\,dx = \tfrac{1}{5}\cos^4 x\sin x + \tfrac{4}{15}\cos^2 x\sin x - \tfrac{2}{3}\sin x + C$$

8.9 Integration by partial fractions

Under this heading we are concerned exclusively with what are called *rational functions*: functions with polynomials in x as the denominator and numerator.

In some cases it is of course possible to solve the integration by other means, but very often the quickest is to use the technique of integration using partial fractions. This is particularly appropriate where the polynomials factorize into a convenient form, as we shall see. Using this method we simplify the rational function to be integrated into a combination of forms which are more easily integrated, often directly, or by using the rule in equation (8.26). A rational function of the form $k/(Ax+B)$ would be easily integrable using this equation.

A prerequisite to the use of the method is that the degree (that is, the maximum power of x) of the numerator should always be *less* than that of the denominator, so that for example, the expression:

$$\frac{x^3+2x-1}{x^2-1}$$

would have to be made suitable by taking the factor (x^2-1) out of the numerator so that we have,

$$\frac{x(x^2-1)+3x-1}{x^2-1} = x+\frac{3x-1}{x^2-1}$$

We may now find the integral of the rational function above by writing:

$$\int x\,dx+\int\frac{3x-1}{x^2-1}dx$$

The first part of this integral should present no problem and the second involves partial fractions. Taking the factors of the denominator (x^2-1) we have $(x+1)(x-1)$, so that,

$$\frac{3x-1}{x^2-1} = \frac{3x-1}{(x+1)(x-1)} \tag{8.33}$$

This final expression can now be factorized into:

$$\frac{3x-1}{(x+1)(x-1)} = \frac{A}{x+1}+\frac{B}{x-1}$$

Multiplying by $(x+1)(x-1)$ we can now say that:

$$A(x-1)+B(x+1) = 3x-1 \tag{8.34}$$

and since the roots for the denominator in equation (8.33) are $+1$ and -1 we can, by substitution of these values into equation (8.34), obtain that $A = 2$ and $B = 1$, so that:

$$\int\frac{3x-1}{x^2-1}dx = \int\frac{2}{x-1}dx+\int\frac{1}{x+1}dx$$

and using equation (8.26) we have:

$$\int\frac{3x-1}{x^2-1}dx = 2\log(x-1)+\log(x+1)$$

$$= \underline{\log[(x-1)^2(x+1)]+C}$$

Therefore the entire integral is:

$$\int \frac{x^3+2x-1}{x^2-1}\mathrm{d}x = \tfrac{1}{2}x^2+\log[(x-1)^2(x+1)]+C$$

Where the denominator will not factorize as above so that roots may be readily obtained, we can develop the partial fraction technique such that a third term, C, enters the expression (not to be confused with the constant term, C, as a result of integration). Imagine we must integrate the rational function:

$$\int \frac{1}{x(x^2+1)}\mathrm{d}x$$

(x^2+1) has no real roots (solution using equation (5.32) will verify this as we end up trying to find the square root of a negative value in b^2-4ac), and so we write:

$$\int \frac{\mathrm{d}x}{x(x^2+1)} = \int \frac{A}{x}\mathrm{d}x + \int \frac{Bx+C}{x^2+1}\mathrm{d}x$$

We thus say that $(Bx+C)x + A(x^2+1) = 1$. One solution is that $x=0$, and substituting this we obtain $A=1$. Equally the coefficients of x^2 and x on either side of the equation must balance, so that since

$$(A+B)x^2+Cx+A = 1$$

$A+B=0$ and therefore $B=-1$ and $C=0$. Thus we have to integrate:

$$\int \frac{\mathrm{d}x}{x} - \int \frac{x}{x^2+1}\mathrm{d}x$$

By substitution in the second integral we have, if $u = x^2+1$,

$$\frac{\mathrm{d}x}{\mathrm{d}u} = \frac{1}{2x}$$

So we have to integrate:

$$\int x\frac{1}{x^2+1}\frac{1}{2x}\mathrm{d}u$$

Now $x = (u-1)^{1/2}$ and thus:

$$\int \frac{x\,\mathrm{d}x}{x(x^2+1)} = \int (u-1)^{\frac{1}{2}}\frac{1}{u}\frac{1}{2(u-1)^{\frac{1}{2}}}\mathrm{d}u$$

$$= \tfrac{1}{2}\int \frac{1}{u}\mathrm{d}u$$

$$= \tfrac{1}{2}\log u = \tfrac{1}{2}\log(x^2+1)$$

Thus the complete integral is given by:

$$\int \frac{dx}{x(x^2+1)} = \log x - \tfrac{1}{2}\log(x^2+1) + C$$

Again a few examples are given below which the reader should verify for himself, and then attempt the problems.

$$\int \frac{x\,dx}{(x+1)(x+2)^2} = \log(x+2) - \log(x+1) - 2(x+2)^{-1}$$

$$\int \frac{x\,dx}{(1+x)^2} = \log(x+1) + (x+1)^{-1}$$

$$\int \frac{(2x-3)\,dx}{(x-1)(x-7)} = \tfrac{11}{6}\log(x-7) + \tfrac{1}{6}\log(x-1)$$

Problems 8.9

8.9.1 $\int \frac{(x+2)\,dx}{(x+1)(x-3)}$

8.9.2 $\int \frac{dx}{(x-3)(x-2)}$

8.9.3 $\int \frac{x\,dx}{x^4-1}$

8.9.4 $\int \frac{x^2+5x+1}{x(x+1)^2}dx$

8.9.5 $\int \frac{x+1}{(1-x^2)^{\frac{1}{2}}}dx$

8.10 Integration techniques: a summary

Very often, as was stressed at the end of section 8.5 integration of any given integrand can be carried out using any one of a number of the techniques we have given here. It has been impossible to give more than a few examples, and the importance to the reader of working through the examples and solving the problems given at the end of the sections cannot be underestimated. The greatest asset in the task of integration is that of experience. After some time the ability to pick out the *quickest* means of integration becomes almost second nature! Equally, the integrals of frequently encountered, but non-standard, forms will soon impress themselves on the mind to an extent sufficient to make 'guessing' of integrals followed by differentiation for verification, a relatively quick process. Having said this it is realized that to the vast majority of geographers even on the physical side, the opportunities or need for frequent integration over the long term are relatively few. Thus to most, a basic knowledge of the main methods by which functions may be integrated is to be considered the ideal, and the basic techniques of integration

Table 8.1 Summary of integration methods.

Form of integrand		Diagnostic features	Technique	Section	Remarks
Standard		see section 8.4	Refer to list of equations (8.9) to (8.25)	8.4	Full list see Dwight (1957)
Trigonometric	1	$\int \cos 2x \, dx$ $\int \sin 2x \, dx$, etc.	Use equations (2.8) to (2.20)	2.9	
Trigonometric	2	$\int \sin^n x \, dx$ or similar, integer n	Use reduction formulae	8.8	Repeat technique as necessary
$\int f(Ax+B)dx$			$\int f(Ax+B)dx = \frac{1}{A}F(Ax+B)$	8.5	
Polynomials	1	contain strings of terms, e.g. x^n, x^{n-1} ax^n, etc.	Use standard equation (8.9)	8.4	
	2	e.g. $\int \frac{3x-1}{x^2-1} dx$	Factorize using partial fractions	8.9	
Product of two terms or more in x		$\int f_1(x)f_2(x)dx$ and $f_1(x) = f_2'(x)$	Substitution: let $u = f_2(x)$ then find dx/du $\int f_1(x)f_2(x)\frac{dx}{du}du$	8.7	Repeat technique as necessary
Product of two terms or more in x		$\int f_1(x)f_2(x)dx$	By parts: Substitute $u = f_1(x)$, $dv = f_2(x)$ $\int u dv = uv - \int v du + C$	8.6	Repeat technique as necessary

are summarized here. When an integrand is encountered the reader should run systematically through the check list in table 8.1 to see if any of the techniques are directly applicable. The table is not exhaustive and on some occasions the reader will have to fall back on trial and error.

8.11 Numerical integration

We are now in a position to consider the 'true' answer to the simple problem we looked at at the beginning of the chapter. We were concerned with trying to calculate the total depth of water infiltrated into a soil whose infiltration rate could be represented by the function $y = 15+5x^{-1/2}$, where y was the rate of infiltration in mm h^{-1} and x was the time in hours. The approximations we arrived at in section 8.2 were 9.53 mm and 10.40 mm over the time period from $x = 0{\cdot}1$ to $x = 0{\cdot}5$. Using the experience now gained from the intervening sections we can now confidently calculate the definite integral as:

$$\int_{0\cdot 1}^{0\cdot 5}(15+5x^{-1/2})\mathrm{d}x = \Big[15x+10x^{-1/2}\Big]_{0\cdot 1}^{0\cdot 5}$$

$$= (7{\cdot}50+7{\cdot}07)-(1{\cdot}50+3{\cdot}16) = 9{\cdot}91 \text{ mm}$$

However, there exist two 'short cuts' to numerical integration of this sort which are of use in the determination of the definite integral of a complicated function. Both are, it is strongly emphasized, only approximations, but as we shall see they are slightly more precise than the original geometric methods. The first of these is the *trapezoidal rule.*

8.12 The trapezoidal rule

The main reason for error in our previous geometric approximations of the area under a curve lay in the *inclusion* of small areas *above* the curve in figure 8.2(b) and the *exclusion* of small areas *below* the curve in figure 8.2(a). In both cases however the shape of these areas was very close to being triangular. The smaller our value for δx then the closer the curved portion of each approached the chord between the two points on the curve. The trapezoidal rule makes use of these small areas, and simplifies the calculation of their area by assuming that they are in fact triangular. The resulting form of the columns is thus trapezoidal. For the purpose of comparison we shall choose the same limits as before and calculate the area using the trapezoidal rule. Figure 8.3 shows only a small portion of the curve of $y = 15+5x^{-1/2}$ shown in figure 8.2. Let us now say that the point at which each x-coordinate intersects the curve has the coordinates (x_0,y_0), (x_1,y_1), (x_2,y_2) and so on. In our example there are five such coordinates, indicating four trapeziums. The distance along the x-axis between each of the x-coordinates is constant, and for the time being we shall call this h. The area of the first trapezium, intersecting the curve at points (x_0,y_0) and (x_1,y_1) is thus given by:

$$hy_0+\tfrac{1}{2}h(y_1-y_0)$$

In our case since the curve has a negative gradient the last expression will be negative, as y_0 is greater than y_1. The area of the second trapezium is similarly:

$$hy_1+\tfrac{1}{2}h(y_2-y_1)$$

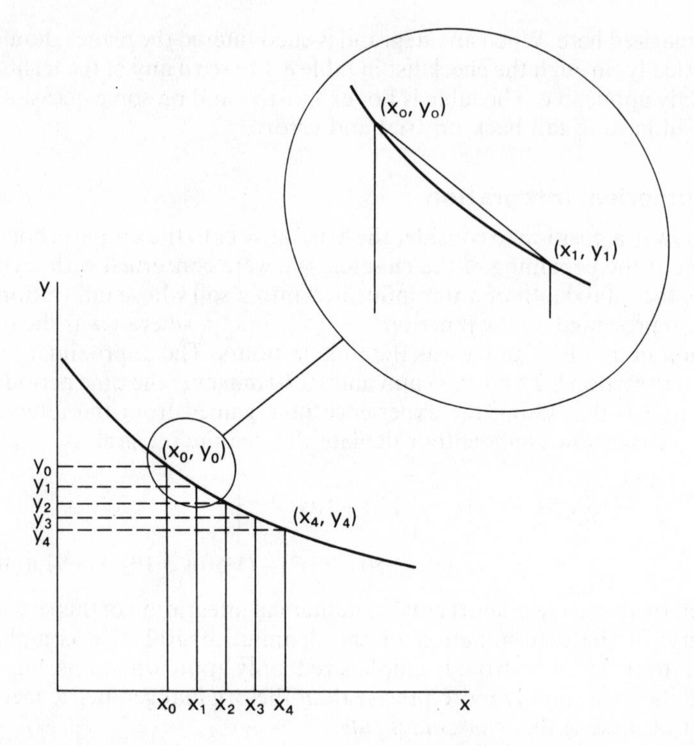

Figure 8.3 Trapezoidal and Simpson's methods of integration.

and so on for the other areas. The total area of all four trapeziums is thus:

$$hy_0+\tfrac{1}{2}h(y_1-y_0)+hy_1+\tfrac{1}{2}h(y_2-y_1)+hy_2+\tfrac{1}{2}h(y_3-y_2)+hy_3+\tfrac{1}{2}h(y_4-y_3)$$

or
$$\tfrac{1}{2}h(y_0+y_1)+\tfrac{1}{2}h(y_1+y_2)+\tfrac{1}{2}h(y_2+y_3)+\tfrac{1}{2}h(y_3+y_4)$$

or
$$\tfrac{1}{2}h[y_0+2(y_1+y_2+y_3)+y_4]$$

Substituting the values for y, where $y_0 = 30{\cdot}8$, $y_1 = 26{\cdot}2$, $y_2 = 24{\cdot}1$, $y_3 = 22{\cdot}9$ and $y_4 = 22{\cdot}1$ we have:

$$\tfrac{1}{2}\times 0{\cdot}1[30{\cdot}8+2(26{\cdot}2+24{\cdot}1+22{\cdot}9)+22{\cdot}1]$$

and thus the trapezoidal solution to the area under the curve between our given limits is 9·97 mm. A general form of the trapezoidal solution is:

$$\int_0^n f(x)\,dx = \tfrac{1}{2}h[y_0+2(y_1+y_2+y_3+\ldots+y_{n-1})+y_n] \tag{8.35}$$

8.13 Simpson's rule

A more accurate approximation still to numerical integration, without resorting to formal integration techniques, is provided by Simpson's rule. This is easily programmed for computer usage and indeed is utilized in many of the program 'packages' now available for numerical analysis. The rule is effectively a more sophisticated, and therefore more precise, version of the trapezoidal rule. Whereas the trapezoidal rule used individual columns in order to calculate area, Simpson's rule makes use of *pairs* of columns. Simpson's rule applies to an *even* number of columns (x_0 to x_5 or x_1 to x_6) and uses Taylor's series, which is derived from Maclaurin's series (see section 6.6), to provide an approximation of the expansion of the function $f(x_n + a)$. If we take the first three x-coordinates (the first two columns) and write $x = x_1 + a$, then we can rewrite the integrand as:

$$\int_{x_0}^{x_2} f(x)\,dx = \int_{-h}^{h} f(x_1 + a)\,da \tag{8.36}$$

between points x_0 and x_2, as the distance between each is h, as in the previous example. Now Taylor's series states that:

$$f(a+x) = f(a) + \frac{f'(a)}{1!}x + \frac{f''(a)}{2!}x^2 + \ldots + \frac{f^{(r)}(a)}{r!}x^r + \ldots$$

and so we can rewrite equation (8.36) as:

$$\int_{-h}^{h} f(x_1 + a)da = \int_{-h}^{h}\left(f(x_1) + \frac{f'(x_1)}{1!}a + \frac{f''(x_1)}{2!}a^2 + \ldots\right)da$$

This integral can be approximated by taking the first three terms (the most significant), such that we have:

$$\left[f(x_1)a + f(x_1)\frac{a^2}{2} + f(x_1)\frac{a^3}{6}\right]_{-h}^{h}$$

$$= 2hf(x_1) + \tfrac{1}{3}h^3 f''(x_1)$$

$$= 2hy_1 + \tfrac{1}{3}h^3 f''(x_1)$$

The second differential in the last term is in fact the *rate of change of the gradient of the tangent to the curve at* x_1, which we can approximate as being about the same as the rate of change of gradient of the *chord* between two adjacent points (as δx tends to dx), so that we now have:

$$f''(x_1) = \frac{1}{h^2}(y_2 - 2y + y_0)$$

and thus the integral becomes:

$$\int_{-h}^{h} f(x_1 + a)\,da = 2hy_1 + \tfrac{1}{3}h(y_2 - 2y_1 + y_0)$$

$$= \tfrac{1}{3}h(y_2 + 4y_1 + y_0)$$

We can repeat this exercise for every pair of columns in the sequence and arrive at a general expression for area, which is Simpson's rule, of:

$$\int_{0}^{n} f(x)dx = \tfrac{1}{3}h[y_0 + 4(y_1 + y_3 + y_5 + \ldots) + 2(y_2 + y_4 + y_6 + \ldots) + y_n] \tag{8.37}$$

In verbal terms, we must therefore take the sum of the *first* and *last* ordinates plus *four* times the sum of the *odd* ordinates plus *twice* the sum of the *even* ordinates. The only condition is that there is an even number of columns. In our particular problem we have such an even number of columns, an odd number of ordinates, and the solution provided by Simpson's rule is:

$$\int_{0\cdot 1}^{0\cdot 5} (15+5x^{-1/2})\,dx = \tfrac{1}{3}\times 0\cdot 1[30\cdot 8+4(26\cdot 2+22\cdot 9)+2(24\cdot 1)+22\cdot 1]$$
$$= 9\cdot 92\,\text{mm}$$

This result is only 0·01 mm greater than the numerical integration performed by formal integration, although clearly it would be very wasteful to use this particular approximation for such a simple integration.

8.14 Applications of integration

We can extend the use of integrals to many different aspects of science and engineering, all of which impinge upon the subject matter of physical geography. Their use is, of necessity, more restricted within physical geography itself, but there is a number of occasions when integrals might be met in the literature. These are far too numerous to mention specifically. The reader will now be aware of the meaning of integration and thus hopefully, be able to interpret them. There are however, two very large fields in which integration is a very important prerequisite. One of these extends the use of integrals from areas, into volumes, centres of gravity and moments of inertia, and the other provides a mathematical basis to *harmonic analysis.*

8.15 Volumes of revolution

Given a simple function relating x and y in two dimensions we can rotate this about either axis to generate a *volume of revolution.* Let us take a very simple case in the first instance. Suppose we wish to generate a cone whose apex lies at the origin of our coordinate system. We can take the general form of the equation of a straight line, $y = mx+k$, and rotate this about the x-axis. The resulting figure for positive values of x, since $k = 0$, will be a cone. The area of the triangle OAB in figure 8.4 is given by the integral:

$$\int_0^{x_1} mx\,dx$$

and this verifies the general formula for the area of a triangle, as the integral is:

$$\left[\tfrac{1}{2}mx^2\right]_0^{x_1} = \tfrac{1}{2}\frac{y_1}{x_1}x_1^2 = \tfrac{1}{2}y_1x_1$$

Each small element of the cone, as we rotate the triangular area, will have the radius mx_n, and so the area of each small element of revolution must be given by $\pi(mx_n)^2$, and thus the volume of revolution for the entire cone is:

$$\pi\int_0^{x_1} (mx)^2\,dx = \tfrac{1}{2}\pi\frac{y_1^2}{x_1^2}x_1^3 = \tfrac{1}{2}\pi y_1^2x_1$$

the area of the base multiplied by half the height.

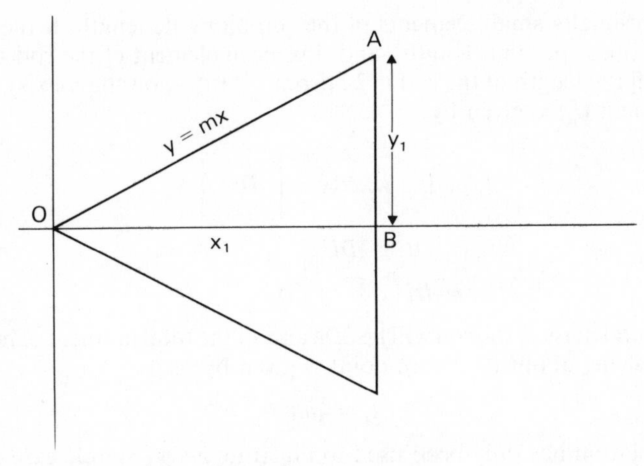

Figure 8.4 Generation of the volume of a cone.

Using this method it is possible to compute the volume of any figure as long as the function representing the form of the side is known. The general expression is:

$$\int_a^b (f(x))^2 \, dx \tag{8.38}$$

For example, the volume of a sphere can be obtained by rotating a circle centred at the origin, about the x-axis. The general equation for a circle is $x^2 + y^2 = r^2$, or $y = (r^2 - x^2)^{\frac{1}{2}}$. The volume of half the sphere can be obtained by finding the integral for the circle between $-r$ and $+r$ (r is the radius), so that we have:

$$\pi \int_{-r}^{+r} (r^2 - x^2) \, dx = \pi \left[r^2 x - \tfrac{1}{3} x^3 \right]_{-r}^{+r}$$

$$= \tfrac{4}{3} \pi^3$$

the usual general expression for the volume of a sphere.

8.16 Moments of inertia

Now let us consider a particle at a point defined on a revolving body. This particle will have a *moment of inertia* about its centre of rotation defined in terms of the mass of the particle (m) and the distance from the centre of revolution (see Davidson 1978). If I is the moment of inertia, and r its distance from the centre of revolution, then:

$$I = mr^2$$

Suppose now that we wish to calculate the *total* moment of inertia for a rod of uniform density, rotating about the y-axis with the axis through its mid-point. Clearly the total moment of inertia for the rod is given by an integration of all the infinitesimally small elements of the rod along its length. If the rod has density D (mass per unit length), and thus each element of the rod has mass $D\mathrm{d}x$, and if the length of the rod is $2\,l$ (from $-l$ to $+l$ on the x-axis), then the total moment (I_t) is given by:

$$I_t = \int_{-l}^{+l} Dx^2\mathrm{d}x = \left[\tfrac{1}{3}Dx^3\right]_{-l}^{+l}$$
$$= \tfrac{1}{3}Dl^3 + \tfrac{1}{3}Dl^3$$
$$= \tfrac{2}{3}Dl^3$$

Now the total mass of the rod will be $2Dl$, and so the total moment of inertia for a rod revolving about its centre point is given by:

$$I_t = \tfrac{1}{3}ml^3$$

This calculation has only been used to illustrate a very simple case in which integration may be used to find the moment of inertia of a body. We could similarly derive moments for a number of other bodies, but this would lie outside the scope of this book. It is left to the individual reader to meet these problems if and when they ever arise.

8.17 Fourier series

The Fourier series is one very useful application of integral calculus when we are considering periodic functions (for example, time series and some types of trend surfaces). We saw in chapter 4 that trigonometric functions, notably those containing sine and cosine, were periodic in that they repeated every 2π radians along the x-axis. These were of course continuous functions, a concept which was defined at the beginning of this chapter. Again, in chapter 6 this time, we saw that functions of trigonometric form could be defined by expansions of series such as Maclaurin's series. Fourier series combine techniques of integral calculus with series expansions, valid for *discontinuous functions.*

Because we are concerned with trigonometric functions whose individual ordinates repeat every 2π radians, we can restrict our attentions to *any* limits containing the interval 2π radians. We shall therefore, for convenience, restrict our terms of reference to integrals over the limits $-\pi < x < +\pi$. Fourier series concern the trigonometric functions $\sin x$, $\sin 2x$, $\sin 3x$, $\sin nx, \ldots$ and $\cos x$, $\cos 2x$, $\cos 3x$, $\cos nx$, ... including constant terms. All such functions regardless of their complexity of terms will have a *period* (wavelength) of 2π. For example, $\sin(2\pi + 3x) = \sin(4\pi + 3x)$. This should be apparent from chapter 2. The basic form of a Fourier series is:

$$f(x) = a_0/2 + a_1 \cos x + a_2 \cos 2x + \ldots + a_r \cos rx + b_1 \sin x + b_2 \sin 2x + b_3 \sin 3x + \ldots + b_r \sin rx \qquad (8.39)$$

This can be simplified to:

$$f(x) = a_0/2 + \sum_{r=1}^{\infty} (a_r \cos rx + b_r \sin rx) \qquad (8.40)$$

where the arbitrary *Fourier coefficients* (a_0, a_r and b_r) are defined by:

$$a_0 = \int_{-\pi}^{\pi} f(x)\,dx \tag{8.41}$$

$$a_r = \int_{-\pi}^{\pi} f(x)\cos rx\,dx \quad (r \text{ is integer}) \tag{8.42}$$

$$b_r = \int_{-\pi}^{\pi} f(x)\sin rx\,dx \quad (r \text{ is integer}) \tag{8.43}$$

In the majority of cases the use of Fourier series yields an approximation to a function of a simpler type. For example, we might envisage the periodic repetition of the function $y = x$ in Fourier terms, such that with the addition of successive terms of higher order r we arrive ultimately at a discontinuous function approximating the line $y = x$ between $x = -\pi$ and $x = +\pi$, when r tends towards infinity. Let us take up this particular example to illustrate the point. To find a_0, a_r and b_r we should have to integrate the relevant functions by substitution into equations (8.41), (8.42) and (8.43). The process is quite long, and there is insufficient space to carry it out here. Integration was by parts and revealed:

$$a_0 = \int_{-\pi}^{\pi} x\,dx = C$$

$$a_r = \int_{-\pi}^{\pi} x\cos rx\,dx = 0$$

$$b_r = \int_{-\pi}^{\pi} x\sin rx\,dx = -\frac{2}{r}\cos\pi r$$

Since $-\cos\pi r = (-1)^{r+1}$, then:

$$b_r = \frac{2(-1)^{r+1}}{r}$$

and thus (from equation (8.40)) our Fourier expansion is:

$$b_r \sin rx = 2\sin x - \sin 2x + \sin 3x - \tfrac{1}{2}\sin 4x + \sin 5x - \tfrac{1}{3}\sin 6x + \ldots \tag{8.44}$$

As we add more and more terms to the series we approach closer and closer the form $y = x$. Note however that the function is discontinuous at $r = \infty$. The graphs drawn in figure 8.5 illustrate the curves for 2, 10 and 50 terms together with that of the function $y = x$.

In physical geography *Fourier analysis* as it is termed, forms an important part of the theory behind *time-series* analysis, and is also an important cornerstone to some aspects of *trend surface* analysis. We shall be following up an example of the former below, but for those wishing to follow up the latter point in geographical literature, Rayner (1972) has applied Fourier series to a study of landform trend surfaces.

The general form of the Fourier series given in equation (8.40) can be modified to fit other limits than $-\pi$ to $+\pi$, whilst preserving the periodic

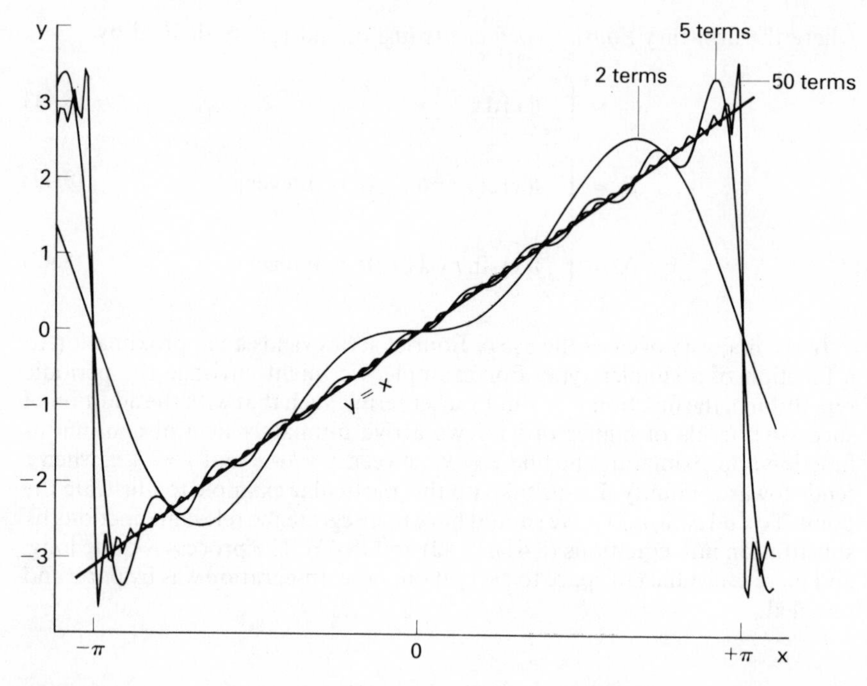

Figure 8.5 $y = x$ and approximations to it using discontinuous harmonic functions of the form in equation (8.40).

nature of the series. The general form of the Fourier series used in harmonic analysis is:

$$f(x) = \bar{x} + \sum_{r=1}^{r=N/2} [a_r \sin(2\pi rx/P) + b_r \cos(2\pi rx/P)] \tag{8.45}$$

where P is the *fundamental period* of the data (see, for example, Panofsky and Brier 1958). So that if we are looking at the annual cycle of monthly rainfall our fundamental period is 12 months. N is the data population and in the case cited above, this too takes the value 12. Note however, that we could still look at the average monthly rainfall retaining $N = 12$, but put $P = 365$ days. The choice of units for our expression obviously conditions the final form of the results of the analysis. The corresponding expressions for a_r and b_r are:

$$a_r = \frac{2}{N}\Big[f(x) \sin(2\pi rx/P)\Big]_{x=1}^{x=N} \tag{8.46}$$

$$b_r = \frac{2}{N}\Big[f(x) \cos(2\pi rx/P)\Big]_{x=1}^{x=N} \tag{8.47}$$

In the example we shall attempt to fit a periodic function of Fourier form to the annual range of monthly rainfall at Nairobi, Kenya. There is a pronounced annual fluctuation of rainfall, and also a distinct seasonal one with the wettest months being April and November (see table 8.2). The first step is to solve a_r

Table 8.2 Average monthly rainfall (mm) for Nairobi, Kenya. Source: Meterological Office, Tables of Temperature, Relative Humidity and Precipitation for the world: Part IV, 1967, HMSO.

J	F	M	A	M	J	J	A	S	O	N	D
38·1	63·5	124·5	210·8	157·5	45·7	15·2	22·9	30·5	53·3	109·2	86·4

and b_r for $r = 1$, the first harmonic, by substitution for the values x (month number, 1 to 12) and $f(x)$ (rainfall for each month) into equations (8.46) and (8.47). The values for the Fourier coefficients for $r = 1$ are:

$$a_1 = \frac{2}{12}\Big[38{\cdot}1 \sin(2\pi/12) + 63{\cdot}5 \sin(4\pi/12) + 124{\cdot}5 \sin(6\pi/12) + \ldots + 86{\cdot}4 \sin(24\pi/12)\Big]$$

and

$$b_1 = \frac{2}{12}\Big[38{\cdot}1 \cos(2\pi/12) + 63{\cdot}5 \cos(4\pi/12) + 124{\cdot}5 \cos(6\pi/12) + \ldots + 86{\cdot}4 \cos(24\pi/12)\Big]$$

When both these computations are complete we obtain:

$$a_1 = 50{\cdot}2 \quad b_1 = -6{\cdot}6$$

The mean monthly rainfall is 79·8 mm and thus the first approximation to the annual series, given by the first harmonic in equation (8.45) is:

$$f(x) = 79{\cdot}8 + 50{\cdot}2 \sin(\pi x/6) - 6{\cdot}6 \cos(\pi x/6) \qquad (8.48)$$

Substituting values of x into this equation we arrive at the first row of figures in table 8.3. By a similar process we can calculate the other Fourier coefficients, if necessary up to the sixth. The second, third and fourth rows of figures in table 8.3 correspond to the cumulative addition of the second, third and fourth harmonics respectively. The resultant equation for the series up to the fourth harmonic is given by:

$$f(x) = 79{\cdot}8 + 50{\cdot}2 \sin(\pi x/6) - 6{\cdot}6 \cos(\pi x/6) - 56{\cdot}4 \sin(\pi x/3) - 6{\cdot}4 \cos(\pi x/3) - 3{\cdot}8 \sin(\pi x/2) + 26{\cdot}3 \cos(\pi x/2) - 5{\cdot}2 \sin(2\pi x/3) - 8{\cdot}0(2\pi x/3) \qquad (8.49)$$

The figures in the fourth row of table 8.3 bear a close resemblance to those in

Table 8.3 Harmonic analysis of Nairobi rainfall. Computed values using equation (8.45) to 1st, 2nd, 3rd and 4th harmonics.

J	F	M	A	M	J	J	A	S	O	N	D
99·2	120·0	130·0	126·6	110·6	86·4	60·4	39·6	29·6	33·0	49·0	73·2
47·2	74·3	136·4	178·7	156·3	80·0	8·3	−6·1	36·0	85·1	94·7	66·8
43·3	48·0	140·2	205·0	152·5	53·7	12·1	20·2	32·2	58·8	98·5	93·1
42·8	56·5	132·2	204·5	161·0	45·7	11·6	28·7	24·2	58·3	107·0	85·1

table 8.2. Thus already we have a good approximation to the original data, and before going any further and calculating fifth and sixth harmonics it is as well to compute the amount of variance already accounted for. The variance accounted for is given by:

$$\frac{a_1^2 + b_1^2}{2s_x^2} \tag{8.50}$$

where s_x is the standard deviation of the raw data. Using this formula we calculate that the first harmonic accounts for 35% of the total variance, the second, 45%, the third, 10% and the fourth only 1%. The total variance explained by the first four harmonics is 91%, and in fact the first two explain 80% of the total. Clearly these are the most important harmonics in the annual march of monthly rainfall, and we see the reason for this when we compute their periods. The first harmonic has a period given by P/r, or 12 months, and the second harmonic has the period six months (12/2). It was stated at the outset that there was a clear annual and seasonal variation in rainfall, the latter being closely correlated with the seasonal march of the zenithal sun.

As the reader will have seen, the use of integration in physical geography occurs over a wide spectrum of topics. Of immediate and obvious use, the examples given in the last section will provide the fundamental mathematical basis upon which the relatively advanced statistical techniques concerned can be built. The more general treatment of integral calculus as such in the earlier sections will in addition, it is hoped, prove useful in helping the reader to interpret some of the mathematics found in more advanced papers dealing with geomorphology, meteorology and so on. If the reader wishes to further his mathematical knowledge of integration beyond the realm of this chapter, then a number of mathematical texts can be recommended, which are particularly useful and clear in their treatment of integral calculus and related topics. These are: Lang (1973) and Knight and Adams (1975a,b).

9
Elementary differential equations

The techniques developed in chapters 5 and 8 can now be applied to the solution of *differential equations.* As with the topic developed in the last chapter (integration) the realm of mathematics covered by this chapter's title would be sufficient to fill a complete text, let alone the solitary chapter in this book. As with the previous chapter the main problem presented to the student is one of the mastery of the different techniques of solution available, and of identifying which types of differential equation may be solved using which technique. For this reason a large number of problems are included to provide the opportunity for perfecting these techniques.

It is difficult to do more than introduce the reader to the most elementary differential equations in this one chapter, in the hope that if he needs to, or if he wishes to, the basis provided here can be used as a firm foundation upon which to build more advanced concepts and techniques. We shall first of all turn our attention to the methods of solution of differential equations and then later, to the provision of a couple of specific examples of their use from within physical geography.

9.1 The basic equation

We have become familiar throughout this book with the mathematical equation relating two or more functions, or variables in the statistical sense. At the same time, we know from chapters 5 and 8 what is meant by differentiation and integration, and have become used to dealing with these in a very simple equation form. If we take a very simple relationship, say, that of the dry adiabatic lapse rate used to illustrate a basic linear function in chapter 3, we see that the temperature lapse itself can be written in the form:

$$\frac{dT}{dz} = -9{\cdot}8\,(^{\circ}\mathrm{C}\quad \mathrm{km}^{-1}) \tag{9.1}$$

This is essentially the starting point for our look at differential equations. We can integrate this very simple differential equation with respect to z and obtain:

$$T = -9{\cdot}8z + C \tag{9.2}$$

with the constant of integration indicating our need for a starting temperature in calculations using the lapse rate; that at ground level in the majority of cases. Differential equations at their most simple are equations such as (9.1), involving not only terms in x and y but also the derivatives, dy/dx, d^2y/dx^2 and so on. For much of this chapter we shall be concerned solely with differential equations which include only the two variables x and y and the derivative of one of these.

In the last chapter we developed some techniques of integration aimed at assessing the definite integral of the simple relationship between infiltration rate over time for a soil. We conveniently called the infiltration *rate* y, forgetting that it is in fact a derivative of the quantity of water with respect to time. If we now call this derivative dy/dx, then to be strictly correct we should write:

$$\frac{dy}{dx} = 15 - 5x^{-1/2} \tag{9.3}$$

so that the integral would be:

$$y = 15x + 10x^{\frac{1}{2}} + C \tag{9.4}$$

We have effectively solved the differential equation given in equation (9.3). Note that our solution is only a general one in that we have left C undefined. Equation (9.4) is thus the *general solution* of the differential equation in (9.3). We could arrive at a *particular solution* by specifying values for x and y. Thus we could say, for example,

$$\frac{dy}{dx} = 15 - 5x^{-1/2} \quad y = 10, x = 0{\cdot}5$$

or

$$\frac{dy}{dx} = 15 - 5x^{-1/2} \quad y(0{\cdot}5) = 10 \tag{9.5}$$

By substituting (0·5, 10) into the general solution in equation (9.4) we arrive at a value for C:

$$10 = 15 \times 0{\cdot}5 + 10 \times 0{\cdot}5^{-1/2} + C$$
$$C = 18{\cdot}1$$

Thus we have a particular solution of:

$$y = 15x + 10x^{1/2} + 18{\cdot}1 \tag{9·6}$$

The values of x and y which have enabled us to provide a particular solution constitute the *boundary condition.* if we define the relationship at $x = 0$ ($y(0)$) then this is called the *initial condition* of the equation.

A similar sequence can be invoked to solve the equations of motion (chapter 5). Using s to indicate distance, v to indicate velocity, a to indicate acceleration and t, time, we can present the equations of motion as successive solutions of differential equations. Thus we can say:

$$\frac{d^2s}{dt^2} = a \tag{9.7}$$

Solving this equation (integrating with respect to t) we have:

$$\frac{ds}{dt} = at + u \tag{9.8}$$

where u is our arbitrary constant of integration, in this case the initial condition where $t = 0$. Solving this equation again by integrating with respect to t we have:

$$s = \tfrac{1}{2}at^2 + ut + k \tag{9.9}$$

Again k is another arbitrary constant, this time representing the distance from the origin at which the body started to move. Since we generally assume for convenience that the body always starts from the origin ($s = 0$, $t = 0$), this constant takes the value zero.

9.2 Order and degree

We are already familiar with first and second derivatives of a function. Equation (9.7) was an equation containing the second derivative of s with respect to t. We thus say that equation (9.7) was a *second-order* differential equation. Equation (9.8) is a *first-order* differential equation. We illustrate below some further more complicated examples of differential equations, with their respective orders in parentheses.

$$2\frac{dy}{dx} - y + 3xy = 0 \qquad \text{(1st)}$$

$$\frac{d^3y}{dx^3} + 2\frac{dy}{dx}3x = y \qquad \text{(3rd)}$$

$$\frac{d^4y}{dx^4} = 4y + x \qquad \text{(4th)}$$

$$y'' - 2y' + 5x = 0 \qquad \text{(2nd)}$$

$$5 - 2y^{(4)} - 3y = 1 \qquad \text{(4th)}$$

The equation takes the order of the highest derivative in each case, irrespective of whether or not any lower-order derivatives also appear. Much of our attention in this chapter is devoted to the solution of first-order differential equations. In principle the techniques for the solution of second- and higher-order differential equations are similar to those for the solution of first-order equations. The solution of such equations involves manipulation of considerable complexity, although the basis laid down in this chapter is still applicable. The enthusiastic reader is urged to read this chapter and thoroughly digest its contents, and then to further his knowledge of the solution of higher-order equations by reference to one of the advanced mathematical texts cited in the suggestions for further reading.

Often confused with order in a differential equation is its degree. The degree of a differential equation refers not to the derivative especially, but to the highest power to which any term in y in the equation is raised. Thus the presence of the terms y^4 or $(y'')^4$ in an equation renders it fourth degree. Here it is important to note:

$$(y'')^3 \neq y^{(6)} \tag{9.10}$$

The left-hand side of this inequality indicates that the second derivative of y is raised to the power three, whilst the right-hand side indicates the sixth derivative of y with respect to x. Equations containing only terms in x of the first degree are termed *linear* differential equations, whilst those of higher degree are non-linear. For the most part we shall be concerned only with first degree equations.

9.3 Ordinary and partial differential equations

All the equations we have so far illustrated in this chapter belong to the group of differential equations known merely, and perhaps unfairly in the light of their usefulness, as *ordinary*. They involve only *one* each of the independent and dependent variables. *Partial* differential equations on the other hand, are equations where more than one independent variable is involved. We have seen ordinary linear equations containing more that two variables in a previous chapter. Consider the simple equation:

$$x = 5y - 2z \tag{9.11}$$

This can be *partially differentiated* to give the derivatives of x with respect to y and of x with respect to z. When we obtain the derivative of x with respect to y we treat all terms involving z as constants, and similarly for the partial differentiation of x with respect to z, treating all y terms as constant. We shall be using partial differentiation from time to time in this chapter, but will not concentrate specifically on the problem except briefly in this section, and shall not attempt the solutions of partial differential equations as such, although such solutions are for the most part relatively straightforward developments of the techniques illustrated in this chapter. Yet again the enthusiastic reader may follow the topic up elsewhere if he so wishes.

The symbols used for partial derivatives differ from those used for ordinary derivatives in a very important way. Whereas we use the letter d to indicate ordinary differentiation, we use the symbol ∂ for partial differentiation, so that the partial derivative of x with respect to y in equation (9.11) is:

$$\frac{\partial x}{\partial y} = 5 \tag{9.12}$$

and the partial derivative of x with respect to z is:

$$\frac{\partial x}{\partial z} = -2 \tag{9.13}$$

We christen this symbol 'partial d'.

As an example of partial differentiation we may consider the following equation:

$$x = 12y^3 - 3yz + 4z^2y^3 \tag{9.14}$$

so that:

$$\frac{\partial x}{\partial y} = 36y^2 - 3z + 12z^2y^2 \tag{9.15}$$

and

$$\frac{\partial^2 x}{\partial y^2} = 72y + 24z^2y \tag{9.16}$$

Further:

$$\frac{\partial x}{\partial z} = -3y + 8zy^3 \tag{9.17}$$

and

$$\frac{\partial^2 x}{\partial z^2} = 8y^3 \tag{9.18}$$

and so on. Note that:

$$\frac{\partial^2 x}{\partial y \partial z} = \frac{\partial}{\partial y}\frac{\partial x}{\partial z} = -3 + 24zy^2 \tag{9.19}$$

and

$$\frac{\partial^2 x}{\partial z \partial y} = \frac{\partial}{\partial z}\frac{\partial x}{\partial y} = -3 + 24zy^2 \tag{9.20}$$

Again in another example, if:

$$x = y \tan z + e^{2y} - z^3 \log y \tag{9.21}$$

then:

$$\frac{\partial x}{\partial y} = \tan z + 2ye^{2y} - z^3/y \tag{9.22}$$

$$\frac{\partial x}{\partial z} = y \sec^2 z - 3z^2 \log y \tag{9.23}$$

$$\frac{\partial^2 x}{\partial y \partial z} = \sec^2 z - 3z^2/y \tag{9.24}$$

Problems 9.3

Find $\partial x/\partial z$, $\partial^2 x/\partial z^2$ and $\partial x/\partial y$, $\partial^2 x/\partial y^2$ for the following:

9.3.1 $x = 5y(z-2) + 10yz$
9.3.2 $x = z \tan y(1 + z \tan y)$
9.3.3 $x = y \log z + 2zy + z \log y$
9.3.4 $x = 2 \sin z \cos y - 3 \tan zy$
9.3.5 $x = 3(z^2 - y^2) + 2z \tan^{-1} y$

9.4 Solution of ordinary first-order differential equations

Since the solution of differential equations by definition involves the establishment of integrals, similar problems beset this task as faced us when we were solving for integrals. These problems will reveal themselves as we progress from the elementary to the more complicated forms of differential equation through this part of the chapter. The 'rules' for solution are summarized in section 9.7.

The simplest type of differential equation is the identity of the form:

$$\frac{dy}{dx} = f(x) \tag{9.25}$$

where we are simply able to integrate both sides with respect to x so that if:

$$\frac{dy}{dx} = \frac{1}{x}$$

then

$$y = \log x + C$$

The techniques here are of course exactly those which we met in chapter 8, and there should be no need to repeat them or set problems for equations of this type.

9.5 Equations involving functions in y

At the next stage in complexity we have equations involving only the derivative, dy/dx, and the dependent variable, y: that is, equations of the form

$$\frac{dy}{dx} = f(y) \tag{9.26}$$

Their solution is only slightly more difficult than those of the previous form involving the derivative and $f(x)$. Consider the equation:

$$\frac{dy}{dx} = y \tag{9.27}$$

We can treat the terms in dy and dx as separate and multiply both sides by dx, so that:

$$dy = y\,dx$$

Clearly it will not help to attempt to integrate y with respect to x, but dividing the equation through by y we obtain:

$$\int \frac{1}{y}dy = \int dx$$

which, integrating the left-hand side with respect to y and the right-hand side with respect to x, is:

$$\log y = x + C \tag{9.28}$$

Note the important inclusion of the constant C in the solution. As the value of this constant is arbitrary for the general solution we could just as validly write:

$$\log y - x + C = 0$$

which although algebraically not the same as the solution of (9.28) since the sign of C is reversed, is perfectly acceptable within our terms of reference. The reader should now attempt the following problems.

Problems 9.5

9.5.1 $\dfrac{dy}{dx} = 5y^2$

9.5.2 $\dfrac{dy}{dx} = \dfrac{4}{y}$

9.5.3 $3\dfrac{dy}{dx} = -2\sin^2 y$

9.5.4 $\dfrac{dy}{dx} = \frac{1}{2}e^{2y}$

9.5.5 $2(4+y^2)\frac{dy}{dx} = 3$

9.6 Separable variable equations

Next we must consider equations containing terms in both x and y, as well as the derivative dy/dx. There is a number of different methods which may be used for their solution, and the choice of method depends upon identifying a general form of equation and then applying the appropriate technique. The simplest technique relates to the simplest and most clear cut of differential equations of this type. Consider the equation:

$$\frac{dy}{dx} = xy \tag{9.29}$$

Multiplying by dx and dividing by y we have:

$$\int \frac{1}{y} dy = \int x \, dx$$

and integrating both sides with respect to y and x respectively:

$$\log y - \tfrac{1}{2}x^2 = C \tag{9.30}$$

In an equation of this form the variables are said to be *separable*, and solution is always possible by this method. We can also illustrate the technique with reference to another example and solve:

$$\frac{dy}{dx} = \frac{2(y-1)}{x+1}$$

Multiplying by dx we have:

$$dy = \frac{2(y-1)}{x+1} dx$$

and $$(x+1)dy = 2(y-1)dx$$

so that:

$$\int \frac{dy}{2(y-1)} = \int \frac{dx}{x+1}$$

and therefore: $$\tfrac{1}{2}\log(y-1) = \log(x+1) + C$$

after integration, or rewriting:

$$\frac{\log(y-1)}{2\log(x+1)} = C$$

The reader is now advised to solve the following problems.

Problems 9.6

9.6.1 $\dfrac{1}{y}\dfrac{dy}{dx} = 5x^3y^2$

9.6.2 $\sin^2 x\dfrac{dy}{dx} - 2y^2 = 0$

9.6.3 $\operatorname{cosec} x\dfrac{dy}{dx} = \sec y$

9.6.4 $x\dfrac{dy}{dx} = 2\operatorname{cosec} y\cos^3 y$

9.6.5 $\dfrac{dy}{dx} = \dfrac{y}{x}\log x$

9.7 Homogeneous equations

A further group of ordinary differential equations can be solved by the substitution of $y = vx$ into the equation, as long as the resulting equation, on substitution, contains terms in only y and v. if this is so, then the equation is said to be a *homogeneous equation.* If $y = vx$, then by using the differentiation of a product rule (equation (5.33)) we have:

$$y = vx$$

$$\frac{dy}{dx} = v\frac{dx}{dx} + x\frac{dv}{dx} = v + x\frac{dv}{dx} \tag{9.31}$$

Let us take as an example, the equation:

$$xy\frac{dy}{dx} = x^2 + y^2 \tag{9.32}$$

$$\frac{dy}{dx} = \frac{x^2 + y^2}{xy} \tag{9.33}$$

Clearly this is not a separable equation. The next stage is thus to determine whether or not the equation is homogeneous. If it is then by substituting $y = vx$ into the right-hand side of the equation, we should obtain an expression containing only terms in v. For equation (9.33) we have:

$$\frac{dy}{dx} = \frac{x^2 + v^2x^2}{vx^2} = \frac{1 + v^2}{v}$$

All terms in x cancel out and thus the equation is homogeneous, so that:

$$\frac{dy}{dx} = f(v)$$

To continue we now equate equation (9.31) to our resulting function of v and get:

$$v + x\frac{dv}{dx} = \frac{1+v^2}{v}$$

This is a separable differential equation and can be solved accordingly:

$$x\frac{dv}{dx} = \frac{1+v^2}{v} - v = \frac{1}{v}$$

$$\therefore \int v\,dv = \int \frac{1}{x}\,dx$$

$$\therefore \tfrac{1}{2}v^2 - \log x = C$$

Now $v = y/x$ and therefore:

$$\underline{y^2 - 2x^2 \log x = 2Cx^2} \qquad (9.34)$$

Now consider an alternative solution for a different value of C. On integrating dx/x above we obtained $\log x + C$. If however we treated the constant itself as a logarithm then our solution would be:

$$\int v\,dv = \int \frac{1}{x}\,dx$$

$$\therefore \tfrac{1}{2}v^2 = \log x + \log C = \log Cx$$

and thus the solution could also be written:

$$\underline{y^2 = 2x^2 \log Cx}$$

Let us now consider a second example:

$$x(y+2x)\frac{dy}{dx} = y(2y+3x)$$

Multiplying out both sides we have:

$$(xy+2x^2)\frac{dy}{dx} = 2y^2 + 3xy$$

or:

$$\frac{dy}{dx} = \frac{2y^2+3xy}{xy+2x^2}$$

Substituting $y = vx$ we have:

$$\frac{dy}{dx} = \frac{2v^2x^2+3vx^2}{vx^2+2x^2}$$

So that, dividing through by x^2 we have:

$$\frac{dy}{dx} = \frac{2v^2+3v}{v+2}$$

or $dy/dx = f(v)$, and the equation is homogeneous. We can now write

$$v + \frac{dv}{dx} = \frac{2v^2 + 3v}{v + 2}$$

and therefore:

$$x\frac{dv}{dx} = \frac{2v^2 + 3v - v(v+2)}{v+2}$$

$$= \frac{v^2 + v}{v + 2}$$

$$\therefore \int \frac{v+2}{v^2+v} dv = \int \frac{1}{x} dx$$

The integral in v may be factorized into:

$$\frac{v+2}{v^2+v} = \frac{A}{v} + \frac{B}{v+1}$$

and thus: $$A(v+1) + Bv = v + 2$$

so that $A = 2$ and $B = -1$. Therefore

$$\int \frac{2}{v} dv - \int \frac{1}{v+1} dv = \int \frac{1}{x} dx$$

$$2 \log v - \log(v+1) = \log Cx$$

Now $v = y/x$, and so:

$$2\log(y/x) - \log(y/x + 1) = \log Cx$$

$$(y/x)^2/(y/x+1) = Cx$$

$$(y/x)^2 = Cx(y/x+1) = C(y+x)$$

$$\underline{y^2 = Cx^2(y+x)}$$

This technique provides a very useful means of solution. Where our differential equation is not separable we should first of all always test for homogeneity by the substitution of $y = vx$. Very few equations are homogeneous. Take as a further example the equation:

$$\frac{dy}{dx} = \frac{x+y}{xy} \qquad (9.35)$$

of a very similar nature to equation (9.32). When we substitute $y = vx$ however, we obtain:

$$\frac{dy}{dx} = \frac{x + vx}{vx^2} = \frac{1+v}{vx}$$

Note that the terms in x do not cancel out and the equation is therefore not homogeneous, and:

$$\frac{dy}{dx} \neq f(v)$$

When this state of affairs arises we must resort to more advanced methods of solution, outlined in sections 9.8 to 9.10.

Problems 9.7

Test the following equations for homogeneity and solve if they satisfy the test:

9.7.1 $\frac{dy}{dx} = \frac{xy - y^2}{x^2}$

9.7.2 $xy^3 \frac{dy}{dx} = x^4 + y^4$

9.7.3 $\frac{dy}{dx} = \frac{2y - x}{2x - y}$

9.8 Solution of 'exact' differential equations

In the previous sections of this chapter we have seen the solution of differential equations which are not only first order and linear, but of a very simple type and whose solution is simply obtained. Where equations occur of a type which is not capable of solution by the methods outlined in sections 9.4 to 9.7 we have to start searching for other ways of approaching the solution. Some first-order equations of a particular type can be solved using the technique outlined here. Others for which the technique cannot be used directly, can frequently be converted into a more appropriate form, and then this technique can be applied. We are here concerned with differential equations which are *exact*.

A differential equation is exact if it takes the form:

$$Q\frac{dy}{dx} + P = 0 \tag{9.36}$$

where P and Q are terms in x and y, and where *the partial derivative of Q with respect to x equals the partial derivative of P with respect to y*, or:

$$\frac{\partial Q}{\partial x} = \frac{\partial P}{\partial y} \tag{9.37}$$

Let us take as illustrative examples the two very similar differential equations:

$$(5x^2 - 2xy)\frac{dy}{dx} + 10xy - y^2 = 0 \tag{9.38}$$

and

$$(5x^3 - 2xy)\frac{dy}{dx} + 10xy - y^2 = 0 \tag{9.39}$$

To the untutored eye the only difference is a relatively unremarkable one, in that the first term $5x^2$ in equation (9.38), becomes $5x^3$ in equation (9.39). However, if we apply the definitions in equations (9.36) and (9.37) we see that although both equations take the form, the partial derivatives of $\partial Q/\partial x$ and $\partial P/\partial y$ do not satisfy the equation (9.37) in the second equation, but do in the first. For equation (9.38) we have:

$$\frac{\partial Q}{\partial x} = 10x - 2y \tag{9.40}$$

and

$$\frac{\partial P}{\partial y} = 10x - 2y \tag{9.41}$$

whilst for equation (9.39):

$$\frac{\partial Q}{\partial x} = 15x^2 - 2y \tag{9.42}$$

and

$$\frac{\partial P}{\partial y} = 10x - 2y \tag{9.43}$$

Clearly the equation (9.38) satisfies the condition in (9.37) whilst equation (9.39) does not. The confirmation that a differential equation is exact considerably eases its solution.

Once we have determined that a differential equation is exact we may proceed as follows. We first define a further function u, in terms of y and x so that:

$$\frac{\mathrm{d}u}{\mathrm{d}x} = \frac{\partial u}{\partial y}\frac{\mathrm{d}y}{\mathrm{d}x} + \frac{\partial u}{\partial x} \tag{9.44}$$

We can see immediately a similarity between this equation and the original general form of the differential equation in (9.36) and so if we write:

$$Q = \frac{\partial u}{\partial y} \quad \text{and} \quad P = \frac{\partial u}{\partial x}$$

equation (9.44) becomes:

$$\frac{\mathrm{d}u}{\mathrm{d}x} = Q\frac{\mathrm{d}y}{\mathrm{d}x} + P = 0 \quad \text{(from equation (9.36))}$$

and thus we may rewrite equation (9.36) as:

$$\frac{\mathrm{d}u}{\mathrm{d}x} = 0$$

or

$$u = \text{constant}$$

We thus have a means of solution of an equation of this form. Let us now solve equation (9.38) in this way:

$$\frac{\partial u}{\partial x} = P = 10xy - y^2 \tag{9.45}$$

$$\frac{\partial u}{\partial y} = Q = 5x^2 - 2xy \tag{9.46}$$

Integrating (9.45) with respect to x we have:

$$u = 5x^2y - y^2x + f(y) + C \tag{9.47}$$

where $f(y)$ is some arbitrary function of y which must be determined in order to arrive at the ultimate solution. Noting that, and noting that we already have a value for $\partial u/\partial y$ (9.46) we can differentiate (9.47) with respect to y, so that:

$$\frac{\partial u}{\partial y} = 5x^2 - 2xy + \frac{df}{dy} \tag{9.48}$$

and equating (9.48) with (9.46) we have:

$$5x^2 - 2xy = 5x^2 - 2xy + \frac{df}{dy}$$

so that

$$\frac{df}{dy} = 0$$

or

$$f(y) = C$$

Thus our solution must be:

$$\underline{5yx^2 - y^2x = C} \tag{9.49}$$

noting that u is a constant and combining this with C.

In the above example $f(y)$ conveniently solved as a constant. Often this does not happen. For example, take the equation:

$$(y^2 + x^2)\frac{dy}{dx} + 2xy = 0 \tag{9.50}$$

so that

$$Q = x^2 + y^2 \quad \text{and} \quad \frac{\partial Q}{\partial x} = 2x \tag{9.51}$$

and

$$P = 2xy \quad \text{and} \quad \frac{\partial P}{\partial y} = 2x \tag{9.52}$$

Now to find the function u, we have:

$$\frac{\partial u}{\partial x} = 2xy \quad \text{and} \quad \frac{\partial u}{\partial y} = x^2 + y^2$$

so that

$$u = x^2y + f(y)$$

Now, differentiating with respect to y we have:

$$\frac{\partial u}{\partial y} = x^2 + \frac{df}{dy}$$

From (9.51) therefore:

$$x^2 + \frac{df}{dy} = x^2 + y^2$$

$$\therefore \frac{df}{dy} = y^2$$

$$\therefore \underline{f(y) = \tfrac{1}{3}y^3 + C}$$

and therefore the solution to (9.50) is:

$$\underline{x^2y + \tfrac{1}{3}y^3 = C}$$

The process of solution of exact equations is thus a simple one of algebraic manipulation. However, as usual, we rarely encounter differential equations which are exact! More often than not we must convert them into an exact form first of all with the use of an *integrating factor*, defined in section 9.9. First of all however, the reader is again urged to attempt the problems below in order to familiarize himself with the technique of solution of exact differential equations.

Problems 9.8

9.8.1 $2(y\sin x+2xy)\dfrac{dy}{dx}+y^2(\cos x+1)=0$

9.8.2 $(\log x+2yx^2+2yx)\dfrac{dy}{dx}+y\left(\dfrac{1}{x}+2yx+y\right)-\log x=0$

9.8.3 $x^2+3y+(3x-y^2)\dfrac{dy}{dx}=0$

9.8.4 $(\sin y-\tan x)\dfrac{dy}{dx}=y\sec^2 x$

9.8.5 $(x-1)\dfrac{dy}{dx}+y-x=1$

9.9 The integrating factor

Consider the equation:

$$2y\cos y+y\left(\frac{1}{x}-x\sin y\right)\frac{dy}{dx}=0 \tag{9.53}$$

This is of the general form shown in equation (9.36) but solving for the partial derivatives of the coefficients we obtain:

$$\frac{\partial Q}{\partial x}=y\log x-y\sin y \tag{9.54}$$

$$\frac{\partial P}{\partial y}=-2y\sin y+2\cos y \tag{9.55}$$

The equation is not therefore exact, and cannot be solved immediately as was outlined in section 9.8. However, if we multiply equation (9.53) by x/y we obtain:

$$2x\cos y+(1-x^2\sin y)\frac{dy}{dx}=0 \tag{9.56}$$

Here, $$\frac{\partial Q}{\partial x}=\frac{\partial P}{\partial y}=-2x\sin y$$

and the factor x/y is known as the *integrating factor*. We may now proceed with the solution as before:

$$\frac{\partial u}{\partial x} = 2x\cos y$$

$$\therefore\ u = x^2\cos y + f(y) + C$$

and $$\frac{\partial u}{\partial y} = -x^2\sin y + \frac{df}{dy}$$

From equation (9.53) we see that:

$$\frac{\partial u}{\partial y} = Q = 1 - x^2\sin y$$

and so $$\frac{df}{dy} = 1 \quad \text{and} \quad f(y) = y + C$$

so our solution is:

$$\underline{x^2\cos y + y = C} \tag{9.57}$$

This was a very simple example where the integrating factor was relatively obvious. As with so many of the techniques involved with integral calculus and the solution of differential equations the task of finding a suitable factor can be an arduous or at least a tedious one, and may ultimately depend upon the chance recognition of equations of a particular form. We are forced yet again to resort to trial and error, or what tends to be termed euphemistically, 'solution by inspection'! Some further problems are set at this point. All the problems in this section may be solved with the application of an integrating factor.

Problems 9.9

9.9.1 Solve

$$(2\tan x + 2x\sec x)\frac{dy}{dx} + y(1+\sec x) = 0$$

by applying the integrating factor $y\cos x$

9.9.2 Solve $(3x^3y^{-1} + 2x^2 - x)\dfrac{dy}{dx} + 6x^2 + yx = 0$

9.9.3 Solve $x\log x\dfrac{dy}{dx} + (x+y) = 0$

9.9.4 Solve $(x^2\sec y - 3xy^2\cot y)\dfrac{dy}{dx} + x(2\operatorname{cosec} y - y^3\cot y) = 0$

9.9.5 Solve $(6x^2y^3 - 18y^2x + 12y)\dfrac{dy}{dx} + 4y^4x - 9y^2 = 0$

9.10 The integrating factor $e^{\int A dx}$

Certain equations of the form:

$$\frac{dy}{dx} + Ay = B \tag{9.58}$$

where A and B are functions of x, can be solved with the substitution of an integrating factor of the form $e^{\int A dx}$, so that:

$$\frac{d}{dx}(e^{\int A dx} y) = e^{\int A dx} B \tag{9.59}$$

For example, consider the equation:

$$x\frac{dy}{dx} = x^3 - y \tag{9.60}$$

This may be put into the form of equation (9.58) simply by dividing through by x and re-arranging the terms:

$$\frac{dy}{dx} + \frac{y}{x} = x^2$$

or

$$\frac{dy}{dx} + \frac{1}{x}y = x^2$$

so that $A = 1/x$ and $B = x^2$,

and: $$e^{\int A dx} = e^{\int (1/x) dx} = e^{\log x} = x$$

so that substituting into equation (9.59) we have:

$$\frac{d}{dx}(xy) = x^3$$

$$\therefore\ xy = \tfrac{1}{4}x^4 + C \quad \text{or} \quad x(4y - x^3) = C$$

Problems 9.10

Identification of the form of differential equations is an important part of their solution. The problems set below involve the solution of differential equations using the techniques outlined in sections 9.8, 9.9 and 9.10. The reader must therefore determine the nature of the equation before applying one of the three techniques to obtain the solution.

9.10.1 $3x\frac{dy}{dx} + 3y = 2$

9.10.2 $(x-1)\frac{dy}{dx} - xy\log x = 2$

9.10.3 $2y\cot x + 2\sin x\frac{dy}{dx} = 0$

9.10.4 $y^2 \cot x(\cot x - \sin y)\dfrac{dy}{dx} + y^3 \sec x \operatorname{cosec} x = 0$

9.10.5 $x^2 + 3y + (3x - y^2)\dfrac{dy}{dx} = 0$

9.11 Summary of solution techniques

The major problem involved in solving first-order differential equations is to determine which basic form the equation takes; for its form in turn dictates our approach to its solution. Once this task is completed a secondary problem is often that of determining integrals of the relevant functions within the equation. At this point in the solution the reader is referred to the summary of integration methods in section 8.10.

As an aid to easy reference the solution techniques for differential equations described in sections 9.4 to 9.10, a similar list of methods is given here. Very broadly, table 9.1 lists equation types in ascending order of complexity. However, with practice and experience 'standard' forms can be recognized relatively easily, obviating the need to work systematically down the list.

Table 9.1 Differential equation solution methods: summary.

Equation form		Method of solution	Section
Simple identity	$\dfrac{dy}{dx} = f(x)$	Integration	9.4
Simple identity	$\dfrac{dy}{dx} = f(y)$	Divide through by $f(y)$ and integrate	9.5
Separable variable	$\dfrac{dy}{dx} = \dfrac{f(y)}{f(x)}$	Divide through as appropriate and integrate	9.6
Homogeneous	$f(xy)\dfrac{dy}{dx} = f(x)\,f(y)$	Substitute $y = vx$ to test for homogeneity. If so, $v + x\dfrac{dv}{dx} = f(v)$ Solve and substitute in for v	9.7
Exact	$Q\dfrac{dy}{dx} + P = 0$, $\dfrac{\delta Q}{\delta x} = \dfrac{\delta P}{\delta y}$	$\delta u/\delta y = P$. Obtain u. Obtain $\delta u/\delta y = Q$, and then $f(y)$ Solution given by u	9.8
Inhomogeneous	$Q\dfrac{dy}{dx} + P = 0$, $\dfrac{\delta Q}{\delta x} = \dfrac{\delta P}{\delta y}$	Attempt to obtain integrating factor so that $\dfrac{\delta Q}{\delta x} = \dfrac{\delta P}{\delta y}$ Then proceed as above	9.9
Inhomogeneous and non-exact	$\dfrac{dy}{dx} + Ay = B$	Write: $\dfrac{d}{dx} e^{\int A dx} y = e^{\int A dx} B$	9.10

9.12 Solution of higher-order differential equations: equations with constant coefficients

We now look at the solution of higher-order differential equations, but restrict our attentions at the outset to those which are both linear and of only second order, although the techniques shown can be equally applied to more complicated higher-order forms. There are three main methods for the solution of higher-order differential equations. The first of these is the general solution for an equation of the form:

$$a_0\frac{d^n y}{dx^n}+a_1\frac{d^{n-1}y}{dx^{n-1}}+\ldots+a_{n-1}\frac{dy}{dx}+a_n y = 0 \qquad (9.61)$$

Since we are restricting ourselves to only second-order equations we can write in a general second-order linear form:

$$a\frac{d^2y}{dx^2}+b\frac{dy}{dx}+cy = 0 \qquad (9.62)$$

This is a homogeneous equation of a form which may be solved by the substitution of $y = e^{mx}$, where m is some arbitrary constant. Since the first derivative of e^{mx} is me^{mx}, and the second derivative is m^2e^{mx} (and so on), then by substitution equation (9.62) becomes:

$$am^2+bm+c = 0$$

as all terms divide through by e^{mx}. This is known as the *auxiliary equation* of the differential equation in (9.62), and we may solve for its roots, m_1 and m_2, by factorizing or substituting into equation (5.32). Our result becomes:

$$y = Ae^{m_1x}+Be^{m_2x} \qquad (9.63)$$

where A and B are arbitrary constants. The number of arbitrary constants in the general solution reflects the order of the original differential equation, as this determines the number of roots we can obtain from the auxiliary equation. So, for a third-order equation the general solution would be:

$$y = Ae^{m_1x}+Be^{m_2x}+Ce^{m_3x} \qquad (9.64)$$

As an illustration of the use of this technique let us take the equation:

$$2\frac{d^2y}{dx^2}+11\frac{dy}{dx}-6y = 0$$

This becomes, by substitution of $y = e^{mx}$,

$$2m^2+11m-6 = 0$$

Factorizing we have:

$$(m+6)(2m-1) = 0$$

or $m_1 = -6$ and $m_2 = \frac{1}{2}$; the general solution is therefore:

$$\underline{y = Ae^{-6x}+Be^{\frac{1}{2}x}}$$

Now consider the situation where $m_1 = m_2$. In the equation below we have such a situation with the root being -1:

$$\frac{d^2y}{dx^2}+2\frac{dy}{dx}+y = 0 \qquad (9.65)$$

Thus the solution is:

$$y = (A+B)e^{-x}$$
$$= Ce^{-x} \quad (\text{where } A+B = C)$$

and we have only *one* arbitrary constant, which does not satisfy the general solution of a second-order differential equation. We must therefore make another, more general solution of $y = ue^{mx}$, where $u = f(x)$, so that substituting in equation (9.65) we have:

$$\frac{d^2}{dx^2}(ue^{-x}) + \frac{d}{dx}(ue^{-x}) + ue^{-x} = 0 \quad (9.66)$$

Using the formula for the differentiation of a product (equation (5.33)) we have:

$$\frac{d}{dx}(ue^{-x}) = -ue^{-x} + e^{-x}\frac{du}{dx} \quad (9.67)$$

and
$$\frac{d^2}{dx^2}(ue^{-x}) = ue^{-x} + -e^{-x}\frac{du}{dx} + e^{-x}\frac{d^2u}{dx^2} + \frac{du}{dx}(-e^{-x}) \quad (9.68)$$

and substituting equations (9.67) and (9.68) into (9.66) we have:

$$ue^{-x} - 2e^{-x}\frac{du}{dx} + e^{-x}\frac{d^2u}{dx^2} + 2\left(e^{-x}\frac{du}{dx} - ue^{-x}\right) + ue^{-x} = 0$$

or
$$\frac{d^2u}{dx^2} = 0$$

$$\therefore \frac{du}{dx} = A$$

$$\underline{u = Ax + B}$$

So the general solution of equation (9.65) must be of the form:

$$\underline{y = (Ax+B)e^{-x}} \quad (9.69)$$

Roots of the auxiliary equation are often imaginary. That is, the equation will not factorize and substitution into equation (5.32) involves the taking of the square root of a negative number. Such solutions are possible, but involve *complex* numbers (those which involve $i = (-1)^{1/2}$), and we are not directly concerned in this book with the integrals of such numbers. The reader should consult a more advanced text in mathematics such as Flanders and Price (1973), if he wishes to follow up this aspect of the solution of differential equations.

Problems 9.12

9.12.1 $2\frac{d^2y}{dx^2} = 8y$

9.12.2 $\frac{d^2y}{dx^2} + 12\frac{dy}{dx} + 36y = 0$

9.12.3 $2\frac{d^2y}{dx^2} - 3\frac{dy}{dx} = 2y$

9.12.4 $5\frac{d^2y}{dx^2} + 14\frac{dy}{dx} - 9y = 0$

9.12.5 $25\frac{d^2y}{dx^2} + 36y = 60\frac{dy}{dx}$

9.13 Particular integrals

The type of equation dealt with in (9.12) is of a very simple nature. Consider the situation if the equation is not homogeneous, so that:

$$a\frac{d^2y}{dx^2} + b\frac{dy}{dx} + cy = f(x) \tag{9.70}$$

We cannot use the technique outlined in (9.12) to solve this equation. However, we can use the *complementary function, u*, to equation (9.70), which is also a function of x, so that we have:

$$a\frac{d^2u}{dx^2} + b\frac{du}{dx} + cu = 0 \tag{9.71}$$

This is also called the *reduced equation*, and the subsequent substitution of $u = e^{mx}$ will enable us to solve it in the same way as we did in section 9.12. The remaining part of the original function is called the *particular integral, v*. Thus for a complete general solution of a non-homogeneous differential equation of the form in (9.70) we can say:

general solution (y) = complementary function (u) + particular integral (v)

where,

$$a\frac{d^2v}{dx^2} + b\frac{dv}{dx} + cv = f(x) \tag{9.72}$$

We can prove this by obtaining the solution of:

$$a\frac{d^2}{dx^2}(u+v) + b\frac{d}{dx}(u+v) + c(u+v)$$

This produces:

$$a\frac{d^2u}{dx^2} + b\frac{du}{dx} + cu + a\frac{d^2v}{dx^2} + b\frac{dv}{dx} + cv = 0 + f(x) \tag{9.73}$$

from equations (9.71) and (9.72).

The main problem with this method is the establishment of the particular integral, v. This is yet again largely a matter of trial and error. In many cases it is very difficult to find a particular integral. We shall deal with this problem in section 9.14, but first provide an example of the use of an elementary particular integral. Consider the differential equation:

$$\frac{d^2y}{dx^2} - 2\frac{dy}{dx} + y = (x+1)^2 \tag{9.74}$$

Our complementary function, u, yields the auxiliary equation:

$$m^2-2m+1=0$$

or $m=1$, so that we have:

$$u=(Ax+B)\mathrm{e}^x$$

To establish the particular integral we can assume that since the right-hand side of the equation (9.74) is a polynomial, it too will be of that form, so that:

$$v=ax^2+bx+c$$

We now have:

$$\frac{\mathrm{d}^2v}{\mathrm{d}x^2}-2\frac{\mathrm{d}v}{\mathrm{d}x}+v=2a-2(2ax+b)+ax^2+bx+c=(x+1)^2$$

Equating coefficients we have:

$$a=1,\ b-4a=2 \text{ and } 2a-2b+c=1$$

so that $a=1$, $b=6$ and $c=11$. So our particular integral is:

$$v=x^2+6x+1 \tag{9.75}$$

and the general solution is:

$$\underline{y=(Ax+B)\mathrm{e}^x+x^2+6x+11} \tag{9.76}$$

This practice can be used to establish certain types of particular integral. Where $f(x)$ was of the form ae^{bx} we can also assume that the particular integral will take the form $a_1\mathrm{e}^{b_1x}$, and so on. Let us now take a more difficult example and attempt a general solution of:

$$\frac{\mathrm{d}^2y}{\mathrm{d}x^2}-2\frac{\mathrm{d}y}{\mathrm{d}x}-y=6x+\sin x$$

Here we may factorize the left-hand side so that we have, in terms of m:

$$m^2-2m-1=(m-1)^2=0$$

and thus,
$$u=(Ax+B)\mathrm{e}^x$$

For the particular integral, v, we may assume it again takes the form governed by the form of the right-hand side of the equation so that:

$$v=(ax+b)+(c\sin x+d\cos x)$$

with the terms in the first set of brackets relating to the first expression, and $(c\sin x+d\cos x)$ relating to the second term, so that as in the previous example:

$$\begin{aligned}\frac{\mathrm{d}^2v}{\mathrm{d}x^2}-2\frac{\mathrm{d}v}{\mathrm{d}x}-v&=\frac{\mathrm{d}^2}{\mathrm{d}x^2}(ax+b+c\sin x+d\cos x)\\&\quad-2\frac{\mathrm{d}}{\mathrm{d}x}(ax+b+c\sin x+d\cos x)\\&\quad-(ax+b+c\sin x+d\cos x)\\&=6x+\sin x\end{aligned}$$

$$\therefore\ (-c\sin x-d\cos x)-2(a+c\cos x-d\sin x)-(ax+b+c\sin x+d\cos x)$$
$$=6x+\sin x$$

Re-arranging terms we have:

$$\sin x(-c+2d-c)+\cos x(-d-2c-d)+x(-a)+(-2a-b) = 6x+\sin x$$

Equating coefficients, we have:

$$-2c+2d = 1, \ -2d-2c+0, \ -a = 6 \text{ and } -2a-b = 0$$
$$\therefore a = -6, \ b = 12, \ c = -\tfrac{1}{4} \text{ and } d = \tfrac{1}{4}$$

Thus our solution is:

$$\underline{y = (Ax+B)e^x - 6x + 12 - \tfrac{1}{4}\sin x + \tfrac{1}{4}\cos x}$$

Some problems are now given which may be solved using the method of particular integrals.

Problems 9.13

Establish particular integrals and solve:

9.13.1 $\dfrac{d^2y}{dx^2} - \dfrac{dy}{dx} = 6y + (x+1)(x-1)$

9.13.2 $2\dfrac{d^2y}{dx^2} - 11\dfrac{dy}{dx} + 12y = (2x+3)^2$

9.13.3 $4\dfrac{d^2y}{dx^2} - 4\dfrac{dy}{dx} + y = e^{2x}$

9.13.4 $\dfrac{d^2y}{dx^2} - 2\dfrac{dy}{dx} + y = 2\sin x$

9.13.5 $\dfrac{d^2y}{dx^2} - 4y = -10\sin 3x$

9.14 D-operators

As a development of the solution of higher-order equations given in sections 9.12 and 9.13 and one which is particularly useful when trial solutions for the particular integral are not immediately apparent, we can use an 'operator' D to indicate derivatives with respect to x, so that:

$$D = \frac{d}{dx}, \ D^2 = \frac{d^2}{dx^2}, \ D^3 = \frac{d^3}{dx^3} \text{ and so on}$$

Thus equation (9.74) may be written:

$$(D^2 - 2D + 1)y = (x+1)^2 \qquad (9.77)$$

These are known as 'D-operators'. Note however that they are not used in the same way as m in the auxiliary equation, and perform a totally different task. Using this notation we can see that the complementary function is given by the equation $(D^2 - 2D + 1)y = 0$, with an auxiliary equation of the same form indicating that $m = 1$, giving us our solution for u. Now, to find the particular integral of the equation above we simply divide both sides by D^2-2D+1, so that:

$$y=\frac{(x+1)^2}{D^2-2D+1}$$
$$=(x+1)^2(D^2-2D+1)^{-1}$$

We can expand the last term by using the binomial expansion (equation (6.14)) to expand $[1+(D^2-2D)]^{-1}$:

$$y=[1-(D^2-2D)+(D^2-2D)^2-(D^2-2D)^3+\ldots](x+1)^2$$

Now using D to 'operate' upon each term in $(x+1)^2$, and noting that any term of higher order than D^2 is zero, we have:

$$y=(x+1)^2-2+4(x+1)+8$$

and so, $$y=x^2+6x+11,$$

as before in equation (9.76), and the solution is again seen to be:

$$\underline{y=(Ax+B)e^x+x^2+6x+11}$$

The D-operator method often provides a very useful means of solving simultaneous differential equations of the linear form. First-order differential equations are relatively easy to solve, but the solution of higher-order simultaneous equations often results in very unwieldy calculations. However, we are presenting here a couple of very simple examples to illustrate the techniques which can be used. Consider the pair of differential equations:

$$2y_1+\frac{dy_1}{dx}+5y_2=0 \qquad (9.78)$$

$$3y_1+5\frac{dy_1}{dx}+y_1=0 \qquad (9.79)$$

Let $D=d/dx$ so that:

$$(D+2)y_1+5y_2=0 \qquad (9.80)$$

and $$(5D+1)y_2+3y_1=0 \qquad (9.81)$$

Multiplying (9.78) by 3 and (9.79) by $(D+2)$ we have:

$$3(D+2)y_1+15y_2=0 \qquad (9.82)$$

$$3(D+2)y_1+(5D+1)(D+2)y_2=0 \qquad (9.83)$$

Subtracting we have:

$$15y_2-(5D+1)(D+2)y_2=0$$

$$\therefore y_2(-5D^2-9D+17)=0 \qquad (9.84)$$

Treating this as an auxiliary equation in m we also have $m_1=-2{\cdot}95$ and $m_2=1{\cdot}15$, so that:

$$\underline{y_2=Ae^{-2{\cdot}25x}+Be^{1{\cdot}15x}}$$

Now substituting for y_2 in equation (9.78) we have:

$$3y_1+5(-2{\cdot}25Ae^{-2{\cdot}25x}+1{\cdot}15Be^{1{\cdot}15x})+Ae^{-2{\cdot}25x}+Be^{1{\cdot}15x}=0$$

$$\therefore\ 3y_1=10{\cdot}25Ae^{-2{\cdot}25x}-6{\cdot}75Be^{1{\cdot}15x}$$

$$\therefore\ \underline{y_1=3{\cdot}42Ae^{-2{\cdot}25x}-2{\cdot}25Be^{1{\cdot}15x}}$$

As a further example let us solve the following equations simultaneously:

$$3\frac{dy_1}{dx}-6y_2=0 \tag{9.85}$$

$$4y_1-\frac{dy_2}{dx}=0 \tag{9.86}$$

Again letting $D = d/dx$ we have:

$$3Dy_1-6y_2=0 \tag{9.87}$$

$$4y_1-Dy_2=0 \tag{9.88}$$

Multiplying (9.87) by 4 and (9.88) by 3D we have:

$$12Dy_1-24y_2=0 \tag{9.89}$$

and

$$12Dy_1-3D^2y_2=0 \tag{9.90}$$

Subtracting (9.90) from (9.89):

$$3D^2y_2-24y_2=0$$

and thus $m_1 = 2{\cdot}83$ and $m_2 = -2{\cdot}83$, so that

$$\underline{y_2=Ae^{2{\cdot}83x}+Be^{-2{\cdot}83x}}$$

Similarly substituting into equation (9.88) we have:

$$4y_1-2{\cdot}83Ae^{2{\cdot}83x}+2{\cdot}83Be^{-2{\cdot}83x}=0$$

$$\therefore\ \underline{y_1=0{\cdot}71(Ae^{2{\cdot}83x}-Be^{-2{\cdot}83x})}$$

Problems 9.14

Use the D-operator method to solve the following differential equations:

9.14.1 $2\frac{d^2y}{dx^2}+3\frac{dy}{dx}-2y=(x+1)(x-1)$

9.14.2 $D^2+2D+1=x$

9.14.3 $\frac{d^2y}{dx^2}-y=x^3+x$

9.15 Applications

The basic groundwork in differential equations has now been laid. The treatment of them has of necessity been brief and elementary. As was observed at the beginning of the chapter the topic generally has entire books devoted to it rather than a mere chapter. Far more could be said on the topic, as indeed it could have been said for the basic calculus introduced throughout this and the earlier chapters. However, the reader is now at a stage when he may begin to glean further, more advanced, mathematical information from texts written for mathematicians and pure scientists rather than merely for physical geographers. As far as this particular chapter is concerned we have taken the subject of differential equations up to an elementary level, such that the reader will have a basic knowledge of what differential equations are and how relatively straightforward ones may be solved. In actual fact the form of differential equations met in physical geography is rarely such that their solution cannot be attempted by one of the techniques listed and explained. Generally, the use of differential equations is restricted to process studies of a theoretical nature, and as a result equations of more than second order are rare. One very large field of research in physical geography in the past decade has been that of studies of hillslope processes and associated soil development. For a long period previous to this many theories of slope development had been put forward largely independent of thorough quantitative treatment of the physical processes involved. The application of mechanics to movement of particles on a slope was given as an illustrative example of resolution of forces in chapter 3. Further development from such a very simple level has resulted in the detailed treatment of slope form and process, such as that by Kirkby (1977) and Young (1972). More generally such topics within geomorphology have been developed by Scheidegger (1970), and a number of models of hillslope development have been applied to specific types of slope and geographical areas (for example, Shaw and Healy 1977). All these publications involve the use of differential equations in their theoretical treatment of the subject.

Equations of a greater complexity do occur in theoretical atmospheric science and in hydraulics (for example, Youngs and Aggelides 1976), but we must take the view that the geographer concerned with such theoretical treatment will be at least of graduate standing, and should be able to refer usefully to texts in higher mathematics—working of course from a basic knowledge gained from reading this book! A more advanced topic of particular relevance and usefulness would be the solution of partial differential equations, requiring much more knowledge of the properties of partial derivatives themselves than we have given (see, for example, Stephenson 1973). In addition the solution of differential equations of a complicated form may be attempted by the use of *Laplace transformations*. As a general guide the enthusiastic reader in one or both these subjects is referred to Stephenson (1973) or Manougian and Northcutt (1973) for further reading. Of these the former is particularly clearly explained, written primarily for scientists rather than mathematicians.

As the final part of this chapter we turn our attention to a more detailed discussion of two simple applications of differential equations to physical geography, both of which lie in theoretical meteorology.

9.16 Wind fields in the atmospheric boundary layer

Frictional drag between the earth and the atmosphere has the effect of reducing wind speeds and providing a deflection to wind direction near to the surface. The most significant reduction in wind speed and change in direction occurs near to the ground itself in what is called the 'boundary layer', arbitrarily defined here as the lowest 500 metres or so, of the atmosphere. Detailed observation of this increase in wind speed with increase in height reveals that it is not a uniform (linear) one, but that *wind shear*, the rate of change of wind speed with height, is greatest near to the ground itself, with the initial condition that when height (z) is zero, so too is wind speed (v). Wind shear thus decreases with increase in height, so that we can write:

$$\frac{\mathrm{d}v}{\mathrm{d}z} \propto \frac{1}{z} \tag{9.91}$$

The constant of proportionality required to make a basic differential equation of this relationship is a function of wind speed and of the form of the ground itself. In practice we use a term comprised of u_* the *frictional velocity*, and k, von Karman's constant (approximately 0·4), but we combine these here for convenience in the constant b, so that $b = u_*/k$, and:

$$\frac{\mathrm{d}v}{\mathrm{d}z} = b\frac{1}{z} \tag{9.92}$$

so that on integration we have:

$$v = b \log z + C \tag{9.93}$$

where C is an arbitrary constant. We thus see that the non-linear increase of wind speed with height in the boundary layer must take the general form of a semi-logarithmic relationship. This provides a very basic start to the construction of a mathematical model of the boundary layer wind field.

9.17 Equations of atmospheric motion

The derivation of the geostrophic and gradient wind equations is a topic which frequently causes more than a slight hiccough to a student's reading (and understanding) in basic meteorology! For this reason no apology is made for including a second example drawn from the field of the atmospheric sciences. In order that any motion may take place in the atmosphere a force must be applied to produce an acceleration of air particles. Air particles, even at the molecular level, have mass, and we define the force applied as the mass of the particle multiplied by the acceleration produced (see Davidson 1978), or:

$$F = ma \tag{9.94}$$

In the atmosphere it is the pressure gradient force which causes an air particle to accelerate in the horizontal plane. Within the earth's atmosphere we can resolve pressure gradient force into the two horizontal directions, and into the vertical direction when considering other forces. If u, v and w are the speeds in each of these directions, x, y and z, we have:

$$F_x = m\frac{\mathrm{d}u}{\mathrm{d}t} \tag{9.95}$$

$$F_y = m\frac{\mathrm{d}v}{\mathrm{d}t} \tag{9.96}$$

$$F_z = m\frac{\mathrm{d}w}{\mathrm{d}t} \tag{9.97}$$

Since pressure is merely a force acting on unit area we can say that pressure (p) acting along x exerts a force of $p\mathrm{d}y\mathrm{d}z$ on the rectangle ABCD in figure 9.1. Assuming there is a pressure gradient acting, x ($\partial p/\partial x$), then the pressure exerts a force of

$$\left(p+\frac{\partial p}{\partial x}\mathrm{d}x\right)\mathrm{d}y\mathrm{d}z$$

on another rectangle at distance $\mathrm{d}x$ downwind (EFGH). Note here the use of the partial derivative for pressure gradient, since here we have pressure gradients resolved into three directions, with x, y and z as independent variables. The difference between the two forces represents the force due to the pressure gradient and thus:

$$F_x = m\left[p\mathrm{d}y\mathrm{d}z-\left(p+\frac{\partial p}{\partial x}\mathrm{d}x\right)\mathrm{d}y\mathrm{d}z\right] \tag{9.98}$$

$$= -m\left(\frac{\partial p}{\partial x}\mathrm{d}x\mathrm{d}y\mathrm{d}z\right) \tag{9.99}$$

Assuming our air particle has unit mass, and since density (ρ) equals mass/volume, then for unit mass:

$$F_x = -\frac{\partial p}{\partial x}\frac{1}{\rho} \tag{9.100}$$

and similarly:

$$F_y = -\frac{\partial p}{\partial x}\frac{1}{\rho} \tag{9.101}$$

We can also derive a similar expression for height, z, but in this case we put in the term for acceleration due to gravity, g, as that which induces the downward increase in atmospheric pressure (vertical pressure gradient), and therefore:

$$F_z = g = -\frac{\partial p}{\partial z}\frac{1}{\rho} \tag{9.102}$$

So, combining equation (9.95) with (9.100) and equation (9.96) with (9.101) we obtain:

$$\frac{\mathrm{d}u}{\mathrm{d}t} = -\frac{\partial p}{\partial x}\frac{1}{\rho} \tag{9.103}$$

$$\frac{\mathrm{d}v}{\mathrm{d}t} = -\frac{\partial p}{\partial y}\frac{1}{\rho} \tag{9.104}$$

We now add the corrresponding terms for the Coriolis force (again per unit mass):

$$C = 2V\Omega\sin\phi = Vf \tag{9.105}$$

where V is the wind speed down the pressure gradient (that is not resolved), Ω the angular velocity of the earth and ϕ is the latitude. The term f is known as the Coriolis parameter. If we now resolve the entire Coriolis force into the two directions along the y- and x-axes with speeds v and u respectively, and remember that whilst the Coriolis force is proportional to the resolved wind

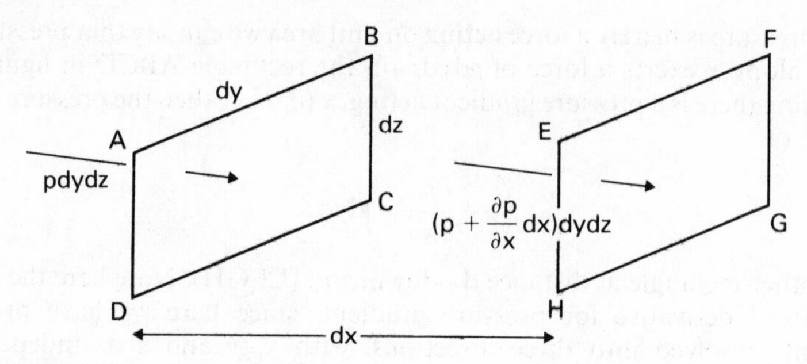

Figure 9.1 Pressure gradient force.

speed, it acts at *right-angles* to it, then we see the component of the Coriolis force, *vf*, has no effect upon dv/dt in the *y*-direction, but does have upon du/dt in the *x*-direction. This is shown diagrammatically in figure 9.2(a) for components in the *x*-directions and in figure 9.2(b) for components in the *y*-direction. Our resulting equations of motion in the *x*- and *y*-directions are thus:

$$\frac{\mathrm{d}u}{\mathrm{d}t} = -\frac{\partial p}{\partial x}\frac{1}{\rho} \tag{9.106}$$

$$\frac{\mathrm{d}v}{\mathrm{d}t} = -\frac{\partial p}{\partial y}\frac{1}{\rho} \tag{9.107}$$

We can develop the equations (9.106) and (9.107) further to show that for conditions where du/dt and dv/dt are zero, a wind is generated which is directly proportional to the pressure gradient, so that for u we have:

$$u = -\frac{\partial p}{\partial y}\frac{1}{f\rho} \tag{9.108}$$

and for v:

$$v = -\frac{\partial p}{\partial x}\frac{1}{f\rho} \tag{9.109}$$

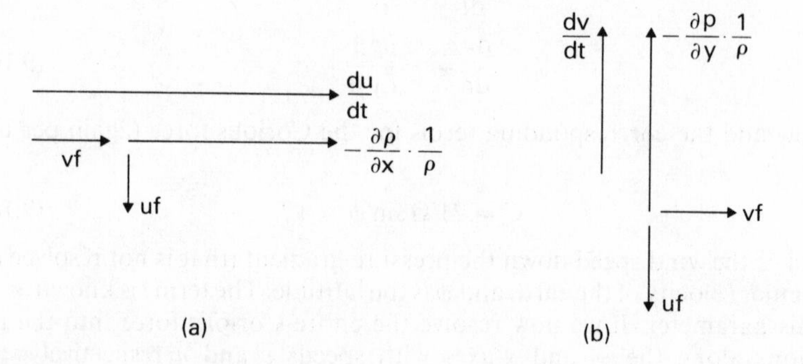

Figure 9.2 Components of Coriolis force.

These conditions, where the acceleration resolved in each direction is zero, are the conditions for *balanced flow*, and the resultant wind generated at right-angles to the pressure gradient (i.e. parallel to the isobars), is the *geostrophic wind*. The wind must blow at right-angles to the pressure gradient since in the expression for u (resolved along x) we have only a pressure gradient term along the y-axis.

We can write a general equation for the geostrophic wind by finding the resultant of u and v at any one time, so that the resultant wind blows parallel to the plotted isobars. Calling this general geostrophic wind V_g we therefore have:

$$V_g = \frac{dp}{dn}\frac{1}{f\rho} \tag{9.110}$$

where n is measured parallel to the isobars. Note that we now use the ordinary differential. Now, for *curved* isobaric flow, we can modify equation (9.110) by adding in the term for the *centripetal acceleration*, the acceleration required to keep the air flowing in a curved path around the isobars. The centripetal acceleration increases with decreasing radius and increasing wind speed, so that, where r is the radius of curvature, we have:

$$fV = \frac{V^2}{r} - \frac{dp}{dr}\frac{1}{\rho} \tag{9.111}$$

for the *anticyclonic* case where the centripetal acceleration acts *against* the pressure gradient force, and:

$$-\frac{V^2}{r} - \frac{dp}{dr}\frac{1}{\rho} = fV \tag{9.112}$$

for the *cyclonic* case where both pressure gradient force and centripetal acceleration act in the same direction.

Epilogue

The average reader who may perhaps have been slightly stretched to understand the last two chapters may well feel that an 'epilogue' is a very suitable end to this book! Some it is true, may well have been able to use the book as a refresher course in mathematics. Those who have taken (and passed!) GCE Advanced Level pure mathematics should have experienced no problems and will fall into this category. To the majority who are not in this group it is hoped that the passage has not been too rough, even if their mathematical background was virtually zero when they started.

However, to both groups the point is very strongly emphasized that to a physical geographer the study of mathematics for its own sake cannot be an end in itself. It is true that it is often possible to derive a mathematical simulation of any number of processes within physical geography, but these must be related to the 'real world' by comparison with observational data using statistical techniques. It is also hoped by the time the reader reaches this point in the book, that when he or she now encounters the formulae describing statistical techniques, which so often seem to deter, these formulae may now be more easily understood. Additionally, in order that mathematical simulation may be attempted by a researcher or understood by the student, he must have a good basic grounding in the physical or chemical processes involved in the phenomena or events being modelled. Frequently the understanding of such processes is considerably eased with reference to mathematical formulae. The gas laws (see Davidson 1978) for example, are more easily remembered in equation form than verbally.

The mathematics developed in this book provide an adequate basis for most undergraduate and many more advanced, needs. The basic physics and chemistry required of physical geographers at undergraduate level may be approached using Davidson's companion book to this volume. As a final word of warning however it is important to stress that as physical geographers we must always consider mathematics as a slave to our needs and never let a preoccupation with mathematics allow it to become a master. Just as it is a servant to scientific thought, method and theories, it must also be a servant to the science of physical geography itself. The identity of physical geography as a subject lies with its being a true science in every way, adopting the rigours and precision of scientific method and of mathematical expression.

Solutions to Problems

Chapter 1

1.6.1	$\bar{2}{\cdot}1761$, $-1{\cdot}8239$
1.6.2	1·0878
1.6.3	1·7547
1.6.4	$\bar{3}{\cdot}3979$, $-2{\cdot}6021$
1.6.5	$\bar{2}{\cdot}7324$, $-1{\cdot}2676$
1.6.6	$\bar{2}{\cdot}7324$, $-1{\cdot}2676$
1.6.7	0·7140
1.6.8	0·0436
1.6.9	1·414, 76,010, 759·7, 1·442

Chapter 2

2.5.1	northerly $-4{\cdot}03$ m/s: westerly 4·03 m/s
2.5.2	2·3: 86 degrees
2.5.3	0·1414 mg
2.5.4	7·7 m/s onshore at 54 degrees to the coastline
2.5.5	5·59 m/s up a slope inclined at $63^\circ 26' 06''$ to the horizontal
2.9.1	$\boldsymbol{a}.\boldsymbol{b} = 27$, $\theta = 34^\circ 45'$; $\boldsymbol{a} \wedge \boldsymbol{b} = -\boldsymbol{i} - 7\boldsymbol{j} + 19\boldsymbol{k}$
2.9.2	$\boldsymbol{a}.\boldsymbol{b} = 0$, $\theta = 90^\circ 00'$; $\boldsymbol{a} \wedge \boldsymbol{b} = -20\boldsymbol{i} - 12\boldsymbol{j} - 4\boldsymbol{k}$
2.9.3	$\boldsymbol{a}.\boldsymbol{b} = -60{\cdot}58$, $\theta = 131^\circ 39'$; $\boldsymbol{a} \wedge \boldsymbol{b} = -37{\cdot}03\boldsymbol{i} - 52{\cdot}07\boldsymbol{j} - 18{\cdot}8\boldsymbol{k}$

Chapter 3

3.2.1	$x = -\frac{5}{9}$, $y = -5$
3.2.2	$x = \frac{1}{4}$, $y = \frac{3}{4}$
3.2.3	$x = \frac{1}{2}$, $y = 1\frac{1}{4}$
3.2.4	$x = \frac{1}{2}$, $y = -\frac{1}{2}$
3.2.5	$x = 2$
3.2.6	$x = 0$, $y = 2$
3.2.7	no solution, lines parallel
3.2.8	$x = 1$, $y = -2$
3.5.1	reduced major axis: $y = 0{\cdot}67x + 0{\cdot}91$ regression x on y: $y = 0{\cdot}68x + 0{\cdot}85$ regression y on x: $y = 0{\cdot}66x + 0{\cdot}96$
3.5.2	reduced major axis: $y = 57{\cdot}7 - 0{\cdot}57x$ regression x on y: $y = 70{\cdot}4 - 0{\cdot}91x$ regression y on x: $y = 49{\cdot}3 - 0{\cdot}35x$

Chapter 4

4.3.1 $\begin{pmatrix} 12 & 19 \\ 16 & 16 \end{pmatrix}$

4.3.2 $\begin{pmatrix} -42 \\ 51 \\ 24 \end{pmatrix}$

4.3.3 $A = B$, $x = 10$, $y = 5$

4.8.1 $|A| = 29$ $\quad \frac{1}{29}\begin{pmatrix} 13 & 11 & 3 \\ 2 & 5 & -4 \\ 18 & 12 & 7 \end{pmatrix}$

4.8.2 $|A| = 1$ $\quad \begin{pmatrix} 6 & -2 & -3 \\ -1 & 1 & 0 \\ -1 & 0 & 1 \end{pmatrix}$

4.8.3 $|A| = 3$ $\quad \begin{pmatrix} \frac{11}{3} & -3 & \frac{1}{3} \\ -\frac{7}{3} & 3 & -\frac{2}{3} \\ \frac{2}{3} & -1 & \frac{1}{3} \end{pmatrix}$

4.10.1 $\lambda = 10{\cdot}75$ or $-0{\cdot}75$; $y = 2{\cdot}2x$, $y = -0{\cdot}7x$
4.10.2 $\lambda = 5$ or 2; $y = -4{\cdot}5x$, $y = -3x$
4.10.3 $\lambda = 3{\cdot}56$ or $-0{\cdot}56$; $y = 0{\cdot}28x$, $y = -1{\cdot}78x$

Chapter 5

5.1.1 $dy/dx = 15x^2$, $d^2y/dx^2 = 30x$
5.1.2 $dy/dx = -4x$, $d^2y/dx^2 = -4$
5.1.3 $dy/dx = x^2 - \frac{2}{3} + \frac{10}{3}x$, $d^2y/dx^2 = 2x + \frac{10}{3}$
5.1.4 $dy/dx = 12$, $d^2y/dx^2 = 0$
5.1.5 $dy/dx = 3x^2 - 1$, $d^2y/dx^2 = 6x$
5.1.6 $dy/dx = -2x^{-3}$, $d^2y/dx^2 = 6x^{-4}$
5.1.7 $dy/dx = -8x^{-5}$, $d^2y/dx^2 = 40x^{-6}$
5.1.8 $dy/dx = 2x - 2$, $d^2y/dx^2 = 2$
5.1.9 $dy/dx = x^2 - 2x$, $d^2y/dx^2 = 2x - 2$
5.1.10 $dy/dx = x^2 - 6x + 9$, $d^2y/dx^2 = 2x - 6$

5.4.1 $(x+2)(x-2)$, $x = \pm 2$
5.4.2 $(x+8)^2$, $x = -8$
5.4.3 $2(x-1)(x+3)$, $x = 1$ or -3
5.4.4 $(3x+1)(x+1)$, $x = -\frac{1}{3}$ or -1
5.4.5 $(1-x)(1+x)$, $x = \pm 1$
5.4.6 $x = -1$ or $\frac{1}{3}$
5.4.7 $x = \pm 2$
5.4.8 $x = 1{\cdot}64$ or $-0{\cdot}24$
5.4.9 $x = \frac{5}{6}$ or $-\frac{1}{2}$
5.4.10 no real roots as $b^2 - 4ac$ is negative
5.4.11 $x = +3$, $+1$ or -3

5.4.12 $x = \frac{1}{5}, \frac{1}{2}$ or -2
5.4.13 (0,0) minimum
5.4.14 (0,4) maximum
5.4.15 (0·19, 0·06) minimum, (−3·52, 49·76) maximum
5.4.16 linear function
5.4.17 (0·58, −0·39) minimum, (−0·58, 0·39) maximum
5.4.18 no stationary points: why?
5.4.19 no stationary points: why?
5.4.20 (1, −1) minimum
5.4.21 $(2, 2\frac{2}{3})$ minimum, (0, 4) maximum
5.4.22 (3, 13) inflexion

5.5.1 $5(4x + \cos x)$
5.5.2 $4(1-4x^2)^{-\frac{1}{2}}$
5.5.3 $6x^{-1}$
5.5.4 $(1-2\log x)x^{-3}$
5.5.5 $2(1-2\sin^2 x)$
5.5.6 $2\tan x \sec^2 x$
5.5.7 $\frac{1}{x}\sec^2(\log x)$
5.5.8 $2e^{2x}[\log(x^2+1)+x(x^2+1)^{-1}]$
5.5.9 $\operatorname{cosec} x(1+\log x - x\log x \cot x)$
5.5.10 $1+\log x - 5\left(\frac{1}{x}\sin x + \frac{1}{x^2}\cos x\right)$

Chapter 6

6.1.1 arithmetic: $a = 2·50$, $d = -1·25$
6.1.2 arithmetic: $a = 2x$, $d = x$
6.1.3 geometric: $a = 6$, $r = 6$
6.1.4 geometric: $a = 2·1$, $r = 4$
6.1.5 geometric: $a = -3$, $r = 0·5$
6.1.6 −6·25, −8·75
6.1.7 $9x$, $11x$
6.1.8 1,679,616: 60,466,176
6.1.9 34,406·4: 550,502·4
6.1.10 −0·024: −0·0058

6.3.1 177·5
6.3.2 −31·2
6.3.3 93
6.3.4 0·8
6.3.5 14

6.4.1 2·5
6.4.2 0·3
6.4.3 7161
6.4.4 5904·8
6.4.5 $0·\dot{6}\dot{0}$

6.5.1 $1+5x+10x^2+10x^3+5x^4+x^5$

6.5.2 $\frac{1}{9}+\frac{2}{27}y+\frac{1}{27}y^2+\frac{4}{243}y^3+\frac{5}{729}y^4$
6.5.3 $625y^4+1000y^3x+600y^2x^2+160yx^3+16x^4$
6.5.4 $2-4x+8x^2-16x^3+32x^4$
6.5.5 $x^4+8x^3+24x^2+32x+16$

Chapter 8

8.4.1 $\frac{5}{3}x^3+\frac{3}{2}x^2-2x+C$
8.4.2 $5(x^3-5x+x^2)+C$
8.4.3 -2
8.4.4 19.30
8.4.5 30 degrees
8.4.6 $-435{\cdot}75$
8.4.7 $1{\cdot}6094$
8.4.8 $\tan x-2\sin x+C$

8.6.1 $\frac{5}{4}x^2(2\log x-1)+C$
8.6.2 $3(x^2+2)\sin x-(x^2+6)x\cos x+C$
8.6.3 $x(\log x-1)+C$
8.6.4 $x(\log x)^2-2x(\log x-1)+C$
8.6.5 $x^3(\log x+\frac{2}{3})+C$

8.7.1 $\frac{5}{2}\sin^2 x+C$
8.7.2 $\frac{1}{2}(\tan^{-1}x)^2+C$
8.7.3 $\frac{2}{5}(1-x)^{\frac{5}{2}}-\frac{2}{3}(1-x)^{\frac{3}{2}}+C$
8.7.4 $\frac{1}{2}\tan^2 x+C$
8.7.5 $-\frac{1}{6}e^{-2x^3}+C$

8.9.1 $\frac{5}{4}\log(x-3)-\frac{1}{4}\log(x+1)+C$
8.9.2 $\log(x-3)-\log(x-2)+C$
8.9.3 $\frac{1}{4}[\log(x+1)+\log(x-1)]-\frac{1}{2}\tan^{-1}x+C$
8.9.4 $\log x-3/(x+1)+C$
8.9.5 $\sin^{-1}x-(1-x^2)^{\frac{1}{2}}+C$

Chapter 9

9.3.1 $\partial x/\partial z=15y$, $\partial^2x/\partial z^2=0$, $\partial x/\partial y=5(3z-1)$, $\partial^2x/\partial y^2=0$
9.3.2 $\partial x/\partial z=\tan y(1+2z\tan y)$, $\partial^2x/\partial z^2=2\tan^2 y$
$\partial x/\partial y=z\sec^2 y(1+2z\tan y)$,
$\partial^2x/\partial y^2=2z\sec^2 y(z+\tan y+3z\tan^2 y)$
9.3.3 $\partial x/\partial z=y/z+2y+\log y$, $\partial^2x/\partial z^2=-y/z^2$
$\partial x/\partial y=\log z+2z+z/y$, $\partial^2x/\partial y^2=-z/y^2$
9.3.4 $\partial x/\partial z=2\cos z\cos y-3y\sec^2 zy$
$\partial^2x/\partial z^2=-2\sin z\cos y-6y^2\tan^2 zy\sec^2 zy$
$\partial x/\partial y=-2\sin z\sin y-3z\sec^2 zy$
$\partial^2x/\partial y^2=-2\sin z\cos y-6z^2\tan zy\sec^2 zy$
9.3.5 $\partial x/\partial z=6z+2\tan^{-1}y$, $\partial^2x/\partial z^2=6$
$\partial x/\partial y=-6y+2z/(1+y^2)$, $\partial^2x/\partial y^2=-6-4yz/(1+y^2)^2$

9.5.1 $5y(x+C)+1=0$
9.5.2 $y^2-8x=C$

9.5.3 $3\cot y - 2x = C$
9.5.4 $x + e^{-2y} = C$
9.5.5 $2y(4 + \frac{1}{3}y^2) - 3x = C$

9.6.1 $2y^2(5x^4 + 4C) + 4 = 0$
9.6.2 $2y(\cot x - C) - 1 = 0$
9.6.3 $\sin y + \cos x = C$
9.6.4 $\frac{1}{2}\tan^2 y - 2\log x = C$
9.6.5 $\log y = \frac{1}{2}(\log x)^2 + C$

9.7.1 $x/y - \log x = C$
9.7.2 $y^4 = 4x^2 \log Cx$
9.7.3 $(y - x) = C(y + x)^3$

9.8.1 $y^2(\sin x + x) + Cy = 0$
9.8.2 $(y - x)\log x + (x + 1)xy^2 - x^2 = C$
9.8.3 $x^3 + 9yx - y^3 = C$
9.8.4 $-y\tan x - \cos y = C$
9.8.5 $y(x - 1) - x(\frac{1}{2}x + 1) = C$

9.9.1 as 9.8.1
9.9.2 $yx(3x + 2y) - \frac{1}{2}y(y + 4x) = C$, (factor y/x)
9.9.3 $x + y\log x = C$, (factor $1/x$)
9.9.4 $2x^2 \sec y - xy(xy^2 - 3xy + 6y) = C$, (factor $\tan x$)
9.9.5 $2y^3x^2 - 9y^2x + 12y = C$, (factor $1/6y$)

9.10.1 $x(3y - 2) = C$
9.10.2 $y + 2x^2 - yx = C$
9.10.3 $2y\sin x = C$
9.10.4 $y\tan x + \cos y = C$
9.10.5 $3xy + \frac{1}{3}(x^3 - y^3) = C$

9.12.1 $y = Ae^{2x} + Be^{-2x}$
9.12.2 $y = (Ax + B)e^{-6x}$
9.12.3 $y = Ae^{-0\cdot 5x} + Be^{2x}$
9.12.4 $y = Ae^{0\cdot 54x} + Be^{-3\cdot 34x}$
9.12.5 $y = (Ax + B)e^{1\cdot 2x}$

9.13.1 $6y = Ae^{-2x} + Be^{3x} - x^2 + 2x - 2$
9.13.2 $y = Ae^{4x} + Be^{1\cdot 5x} + 0\cdot 33x^2 + 1\cdot 11x + 1\cdot 66$
9.13.3 $y = (Ax + B)e^{-\frac{1}{2}x} + e^{2x}/9$
9.13.4 $y = (Ax + B)e^{x} + \cos x$
9.13.5 $13y = Ae^{2x} + Bx^{-2x} + 10\sin 3x$

9.14.1 $2y = Ae^{-2x} + Be^{\frac{1}{2}x} + x^2 + 3x + 11/2$
9.14.2 $y = (Ax + B)e^{-x} + x - 2$
9.14.3 see 9.13.2

Appendix 1

The Greek alphabet and some of the commoner uses of the letters

Letter		English pronunciation	Use in geography
Α	α	alpha	often used as a miscellaneous constant
Β	β	beta	often used as miscellaneous constant
Γ	γ	gamma	often used as a miscellaneous constant
Δ	δ	delta	used in calculus = 'small change in'
Ε	ε	epislon	phase angle (wavelengths)
Ζ	ζ	zeta	= vorticity
Η	η	eta	
Θ	θ	theta	represents angle
Ι	ι	iota	
Κ	κ	kappa	used in Pearson distributions
Λ	λ	lambda	= wavelength
Μ	μ	mu	= micron (10^{-6}m)
Ν	ν	nu	= frequency (of wave phenomena)
Ξ	ξ	xi	
Ο	ο	omicron	
Π	π	pi	used in circular calculations (= 3·1412)
Ρ	ρ	rho	used to represent density
Σ	σ	sigma	used in statistics = summation
Τ	τ	tau	used in statistics (Kendall's T)
Υ	υ	upsilon	
Φ	ϕ	phi	represents angles
Χ	χ	chi	used in statistics (chi-squared test)
Ψ	ψ	psi	
Ω	ω	omega	= angular velocity

Appendix 2

The International System of Units (SI)

A.2.1 Standard units

The following units are in accordance with the recently adopted International System of units. A complete description may be found in '*The International System of Units*', National Physical Laboratory, HMSO, 1970. Many texts in physical geography give values for certain constants in other forms of units which can be confusing. The SI system uses the base units of metre (m) for length, kilogram (kg) for mass, and second (s) for time, whilst the preceeding CGS system used centimetre (cm), gram (g) and second. The British (Imperial) system, now mostly obsolete, except in the USA, used feet, pounds and seconds (FPS). Where possible in this book SI units have been adhered to, except where the citing of research findings has been in other units originally and conversion might lead to confusion.

There are three types of units: *base units, derived units* and *supplementary units.*

A.2.1.1 Base units

Length: metre (m)
the length equal to 1,650,763·73 wavelengths in vacuum of the radiation corresponding to the transition between the levels $2p_{10}$ and $5d_5$ of the krypton-86 atom.

Mass: kilogram (kg)
an international prototype made of platinum–iridium is kept at the BIPM (International Bureau of Weights and Measures).

Time: second (s)
originally 1/86,400 of the mean solar day. Now defined in terms of radiation emission from the caesium-133 atom.

Temperature: Kelvin (K)
the kelvin, the unit of thermodynamic temperature, is the fraction 1/273·16 of the thermodynamic temperature of the triple point of water. (0 °C = 273·16 K)

Luminous intensity: candela (cd)
the candela is the luminous intensity, in the perpendicular direction, of a surface of 1/600,000 square metre of a black body at the temperature of freezing platinum under a pressure of 101,325 newtons per square metre.

Amount of substance: mole (mol)
the mole is the amount of substance of a system which contains as many elementary entities as there are atoms in 0·012 kg of carbon 12.

A.2.1.2 Derived and supplementary units

Area	square metre	m^2
Volume	cubic metre	m^3
Speed, velocity	metres per second	m/s
Acceleration	metres per second per second	m/s^2
Wavenumber	1 per metre	1/m
Density	kilogram per cubic metre	kg/m^3
Luminance	candela per square metre	cd/m^2
Frequency	hertz (Hz)	s^{-1}
Force	newton (N)	$m\ kg\ s^{-2}$
Pressure	pascal (Pa)	N/m^2
Energy	Joule (J)	N/m
Power	watt (W)	J/s

A.2.1.3 Units used with SI system

Degree (angle)	(π/180) radians	(π=3·1412)
Minute (angle)	1/60 degree	
Second (angle)	1/60 minute	
Litre (l)	one cubic decimetre	($10^{-1}\ m^3$)
Tonne (t)	10^3 kg	
Moment of force	newton metre	Nm
Surface tension	newton per metre	N/m
Heat flux density	watt per square metre	W/m^2
Heat capacity	joule per kelvin	J/K
Specific heat	joule per kilogram kelvin	J/kg K
Thermal conductivity	watt per metre kelvin	W/m K
Plane angle	radian (rad)	
Solid angle	steradian (sr)	

(The radian is the plane angle between two radii of a circle which cut off on the circumference an arc equal in length to the radius)

Angular velocity	radian per second	rad/s
Angular acceleration	radian per second per second	rad/s^2

A.2.2 Decimal multiples and submultiples

Prefixes exist for decimal multiples and submultiples of SI units from 10^{12} to 10^{-18}.

10^{12}	tera	T	10^{-1}	deci	d
10^9	giga	G	10^{-2}	centi	c
10^6	mega	M	10^{-3}	milli	m
10^3	kilo	k	10^{-6}	micro	μ
10^2	hecto	h	10^{-9}	nano	n
10^1	deca	da	10^{-12}	pico	p
			10^{-15}	femto	f
			10^{-18}	atto	a

A.2.3 Non-standard units

ångström	(Å)	= 0·1 nm=10^{-10} m
bar	(bar)	= 0·1 MPa=10^{5}Pa (1 millibar=1/1000 bar)
erg	(erg)	= 10^{-7}J
dyne	(dyn)	= 10^{-5}N
poise	(P)	= 0·1 P s
calorie	(cal)	= 4·1868J
micron	(μ)	= 10^{-6}m

Appendix 3

Tables of Four-Figure Logarithms

LOGARITHMS

	0	1	2	3	4	5	6	7	8	9	1	2	3	4	5	6	7	8	9
10	0000	0043	0086	0128	0170	0212	0253	0294	0334	0374	4	8	12	17	21	25	29	33	37
11	0414	0453	0492	0531	0569	0607	0645	0682	0719	0755	4	8	11	15	19	23	26	30	34
12	0792	0828	0864	0899	0934	0969	1004	1038	1072	1106	3	7	10	14	17	21	24	28	31
13	1139	1173	1206	1239	1271	1303	1335	1367	1399	1430	3	6	10	13	16	19	23	26	29
14	1461	1492	1523	1553	1584	1614	1644	1673	1703	1732	3	6	9	12	15	18	21	24	27
15	1761	1790	1818	1847	1875	1903	1931	1959	1987	2014	3	6	8	11	14	17	20	22	25
16	2041	2068	2095	2122	2148	2175	2201	2227	2253	2279	3	5	8	11	13	16	18	21	24
17	2304	2330	2355	2380	2405	2430	2455	2480	2504	2529	2	5	7	10	12	15	17	20	22
18	2553	2577	2601	2625	2648	2672	2695	2718	2742	2765	2	5	7	9	12	14	16	19	21
19	2788	2810	2833	2856	2878	2900	2923	2945	2967	2989	2	4	7	9	11	13	16	18	20
20	3010	3032	3054	3075	3096	3118	3139	3160	3181	3201	2	4	6	8	11	13	15	17	19
21	3222	3243	3263	3284	3304	3324	3345	3365	3385	3404	2	4	6	8	10	12	14	16	18
22	3424	3444	3464	3483	3502	3522	3541	3560	3579	3598	2	4	6	8	10	12	14	15	17
23	3617	3636	3655	3674	3692	3711	3729	3747	3766	3784	2	4	6	7	9	11	13	15	17
24	3802	3820	3838	3856	3874	3892	3909	3927	3945	3962	2	4	5	7	9	11	12	14	16
25	3979	3997	4014	4031	4048	4065	4082	4099	4116	4133	2	3	5	7	9	10	12	14	15
26	4150	4166	4183	4200	4216	4232	4249	4265	4281	4298	2	3	5	7	8	10	11	13	15
27	4314	4330	4346	4362	4378	4393	4409	4425	4440	4456	2	3	5	6	8	9	11	13	14
28	4472	4487	4502	4518	4533	4548	4564	4579	4594	4609	2	3	5	6	8	9	11	12	14
29	4624	4639	4654	4669	4683	4698	4713	4728	4742	4757	1	3	4	6	7	9	10	12	13
30	4771	4786	4800	4814	4829	4843	4857	4871	4886	4900	1	3	4	6	7	9	10	11	13
31	4914	4928	4942	4955	4969	4983	4997	5011	5024	5038	1	3	4	6	7	8	10	11	12
32	5051	5065	5079	5092	5105	5119	5132	5145	5159	5172	1	3	4	5	7	8	9	11	12
33	5185	5198	5211	5224	5237	5250	5263	5276	5289	5302	1	3	4	5	6	8	9	10	12
34	5315	5328	5340	5353	5366	5378	5391	5403	5416	5428	1	3	4	5	6	8	9	10	11
35	5441	5453	5465	5478	5490	5502	5514	5527	5539	5551	1	2	4	5	6	7	9	10	11
36	5563	5575	5587	5599	5611	5623	5635	5647	5658	5670	1	2	4	5	6	7	8	10	11
37	5682	5694	5705	5717	5729	5740	5752	5763	5775	5786	1	2	3	5	6	7	8	9	10
38	5798	5809	5821	5832	5843	5855	5866	5877	5888	5899	1	2	3	5	6	7	8	9	10
39	5911	5922	5933	5944	5955	5966	5977	5988	5999	6010	1	2	3	4	5	7	8	9	10
40	6021	6031	6042	6053	6064	6075	6085	6096	6107	6117	1	2	3	4	5	6	8	9	10
41	6128	6138	6149	6160	6170	6180	6191	6201	6212	6222	1	2	3	4	5	6	7	8	9
42	6232	6243	6253	6263	6274	6284	6294	6304	6314	6325	1	2	3	4	5	6	7	8	9
43	6335	6345	6355	6365	6375	6385	6395	6405	6415	6425	1	2	3	4	5	6	7	8	9
44	6435	6444	6454	6464	6474	6484	6493	6503	6513	6522	1	2	3	4	5	6	7	8	9
45	6532	6542	6551	6561	6571	6580	6590	6599	6609	6618	1	2	3	4	5	6	7	8	9
46	6628	6637	6646	6656	6665	6675	6684	6693	6702	6712	1	2	3	4	5	6	7	7	8
47	6721	6730	6739	6749	6758	6767	6776	6785	6794	6803	1	2	3	4	5	5	6	7	8
48	6812	6821	6830	6839	6848	6857	6866	6875	6884	6893	1	2	3	4	4	5	6	7	8
49	6902	6911	6920	6928	6937	6946	6955	6964	6972	6981	1	2	3	4	4	5	6	7	8
50	6990	6998	7007	7016	7024	7033	7042	7050	7059	7067	1	2	3	3	4	5	6	7	8
51	7076	7084	7093	7101	7110	7118	7126	7135	7143	7152	1	2	3	3	4	5	6	7	8
52	7160	7168	7177	7185	7193	7202	7210	7218	7226	7235	1	2	2	3	4	5	6	7	7
53	7243	7251	7259	7267	7275	7284	7292	7300	7308	7316	1	2	2	3	4	5	6	6	7
54	7324	7332	7340	7348	7356	7364	7372	7380	7388	7396	1	2	2	3	4	5	6	6	7

LOGARITHMS

	0	1	2	3	4	5	6	7	8	9	1	2	3	4	5	6	7	8	9
55	7404	7412	7419	7427	7435	7443	7451	7459	7466	7474	1	2	2	3	4	5	5	6	7
56	7482	7490	7497	7505	7513	7520	7528	7536	7543	7551	1	2	2	3	4	5	5	6	7
57	7559	7566	7574	7582	7589	7597	7604	7612	7619	7627	1	2	2	3	4	5	5	6	7
58	7634	7642	7649	7657	7664	7672	7679	7686	7694	7701	1	1	2	3	4	4	5	6	7
59	7709	7716	7723	7731	7738	7745	7752	7760	7767	7774	1	1	2	3	4	4	5	6	7
60	7782	7789	7796	7803	7810	7818	7825	7832	7839	7846	1	1	2	3	4	4	5	6	6
61	7853	7860	7868	7875	7882	7889	7896	7903	7910	7917	1	1	2	3	4	4	5	6	6
62	7924	7931	7938	7945	7952	7959	7966	7973	7980	7987	1	1	2	3	3	4	5	6	6
63	7993	8000	8007	8014	8021	8028	8035	8041	8048	8055	1	1	2	3	3	4	5	5	6
64	8062	8069	8075	8082	8089	8096	8102	8109	8116	8122	1	1	2	3	3	4	5	5	6
65	8129	8136	8142	8149	8156	8162	8169	8176	8182	8189	1	1	2	3	3	4	5	5	6
66	8195	8202	8209	8215	8222	8228	8235	8241	8248	8254	1	1	2	3	3	4	5	5	6
67	8261	8267	8274	8280	8287	8293	8299	8306	8312	8319	1	1	2	3	3	4	5	5	6
68	8325	8331	8338	8344	8351	8357	8363	8370	8376	8382	1	1	2	3	3	4	4	5	6
69	8388	8395	8401	8407	8414	8420	8426	8432	8439	8445	1	1	2	2	3	4	4	5	6
70	8451	8457	8463	8470	8476	8482	8488	8494	8500	8506	1	1	2	2	3	4	4	5	6
71	8513	8519	8525	8531	8537	8543	8549	8555	8561	8567	1	1	2	2	3	4	4	5	5
72	8573	8579	8585	8591	8597	8603	8609	8615	8621	8627	1	1	2	2	3	4	4	5	5
73	8633	8639	8645	8651	8657	8663	8669	8675	8681	8686	1	1	2	2	3	4	4	5	5
74	8692	8698	8704	8710	8716	8722	8727	8733	8739	8745	1	1	2	2	3	4	4	5	5
75	8751	8756	8762	8768	8774	8779	8785	8791	8797	8802	1	1	2	2	3	3	4	5	5
76	8808	8814	8820	8825	8831	8837	8842	8848	8854	8859	1	1	2	2	3	3	4	5	5
77	8865	8871	8876	8882	8887	8893	8899	8904	8910	8915	1	1	2	2	3	3	4	4	5
78	8921	8927	8932	8938	8943	8949	8954	8960	8965	8971	1	1	2	2	3	3	4	4	5
79	8976	8982	8987	8993	8998	9004	9009	9015	9020	9025	1	1	2	2	3	3	4	4	5
80	9031	9036	9042	9047	9053	9058	9063	9069	9074	9079	1	1	2	2	3	3	4	4	5
81	9085	9090	9096	9101	9106	9112	9117	9122	9128	9133	1	1	2	2	3	3	4	4	5
82	9138	9143	9149	9154	9159	9165	9170	9175	9180	9186	1	1	2	2	3	3	4	4	5
83	9191	9196	9201	9206	9212	9217	9222	9227	9232	9238	1	1	2	2	3	3	4	4	5
84	9243	9248	9253	9258	9263	9269	9274	9279	9284	9289	1	1	2	2	3	3	4	4	5
85	9294	9299	9304	9309	9315	9320	9325	9330	9335	9340	1	1	2	2	3	3	4	4	5
86	9345	9350	9355	9360	9365	9370	9375	9380	9385	9390	1	1	2	2	3	3	4	4	5
87	9395	9400	9405	9410	9415	9420	9425	9430	9435	9440	0	1	1	2	2	3	3	4	4
88	9445	9450	9455	9460	9465	9469	9474	9479	9484	9489	0	1	1	2	2	3	3	4	4
89	9494	9499	9504	9509	9513	9518	9523	9528	9533	9538	0	1	1	2	2	3	3	4	4
90	9542	9547	9552	9557	9562	9566	9571	9576	9581	9586	0	1	1	2	2	3	3	4	4
91	9590	9595	9600	9605	9609	9614	9619	9624	9628	9633	0	1	1	2	2	3	3	4	4
92	9638	9643	9647	9652	9657	9661	9666	9671	9675	9680	0	1	1	2	2	3	3	4	4
93	9685	9689	9694	9699	9703	9708	9713	9717	9722	9727	0	1	1	2	2	3	3	4	4
94	9731	9736	9741	9745	9750	9754	9759	9763	9768	9773	0	1	1	2	2	3	3	4	4
95	9777	9782	9786	9791	9795	9800	9805	9809	9814	9818	0	1	1	2	2	3	3	4	4
96	9823	9827	9832	9836	9841	9845	9850	9854	9859	9863	0	1	1	2	2	3	3	4	4
97	9868	9872	9877	9881	9886	9890	9894	9899	9903	9908	0	1	1	2	2	3	3	4	4
98	9912	9917	9921	9926	9930	9934	9939	9943	9948	9952	0	1	1	2	2	3	3	4	4
99	9956	9961	9965	9969	9974	9978	9983	9987	9991	9996	0	1	1	2	2	3	3	3	4

Appendix 4

Tables of Sine, Cosine and Tangent

NATURAL SINES

°	0′	6′	12′	18′	24′	30′	36′	42′	48′	54′	1′	2′	3′	4′	5′
0	·0000	0017	0035	0052	0070	0087	0105	0122	0140	0157	3	6	9	12	15
1	·0175	0192	0209	0227	0244	0262	0279	0297	0314	0332	3	6	9	12	15
2	·0349	0366	0384	0401	0419	0436	0454	0471	0488	0506	3	6	9	12	15
3	·0523	0541	0558	0576	0593	0610	0628	0645	0663	0680	3	6	9	12	15
4	·0698	0715	0732	0750	0767	0785	0802	0819	0837	0854	3	6	9	12	14
5	·0872	0889	0906	0924	0941	0958	0976	0993	1011	1028	3	6	9	12	14
6	·1045	1063	1080	1097	1115	1132	1149	1167	1184	1201	3	6	9	12	14
7	·1219	1236	1253	1271	1288	1305	1323	1340	1357	1374	3	6	9	12	14
8	·1392	1409	1426	1444	1461	1478	1495	1513	1530	1547	3	6	9	12	14
9	·1564	1582	1599	1616	1633	1650	1668	1685	1702	1719	3	6	9	11	14
10	·1736	1754	1771	1788	1805	1822	1840	1857	1874	1891	3	6	9	11	14
11	·1908	1925	1942	1959	1977	1994	2011	2028	2045	2062	3	6	9	11	14
12	·2079	2096	2113	2130	2147	2164	2181	2198	2215	2233	3	6	9	11	14
13	·2250	2267	2284	2300	2317	2334	2351	2368	2385	2402	3	6	8	11	14
14	·2419	2436	2453	2470	2487	2504	2521	2538	2554	2571	3	6	8	11	14
15	·2588	2605	2622	2639	2656	2672	2689	2706	2723	2740	3	6	8	11	14
16	·2756	2773	2790	2807	2823	2840	2857	2874	2890	2907	3	6	8	11	14
17	·2924	2940	2957	2974	2990	3007	3024	3040	3057	3074	3	6	8	11	14
18	·3090	3107	3123	3140	3156	3173	3190	3206	3223	3239	3	6	8	11	14
19	·3256	3272	3289	3305	3322	3338	3355	3371	3387	3404	3	5	8	11	14
20	·3420	3437	3453	3469	3486	3502	3518	3535	3551	3567	3	5	8	11	14
21	·3584	3600	3616	3633	3649	3665	3681	3697	3714	3730	3	5	8	11	14
22	·3746	3762	3778	3795	3811	3827	3843	3859	3875	3891	3	5	8	11	13
23	·3907	3923	3939	3955	3971	3987	4003	4019	4035	4051	3	5	8	11	13
24	·4067	4083	4099	4115	4131	4147	4163	4179	4195	4210	3	5	8	11	13
25	·4226	4242	4258	4274	4289	4305	4321	4337	4352	4368	3	5	8	11	13
26	·4384	4399	4415	4431	4446	4462	4478	4493	4509	4524	3	5	8	10	13
27	·4540	4555	4571	4586	4602	4617	4633	4648	4664	4679	3	5	8	10	13
28	·4695	4710	4726	4741	4756	4772	4787	4802	4818	4833	3	5	8	10	13
29	·4848	4863	4879	4894	4909	4924	4939	4955	4970	4985	3	5	8	10	13
30	·5000	5015	5030	5045	5060	5075	5090	5105	5120	5135	3	5	8	10	13
31	·5150	5165	5180	5195	5210	5225	5240	5255	5270	5284	2	5	7	10	12
32	·5299	5314	5329	5344	5358	5373	5388	5402	5417	5432	2	5	7	10	12
33	·5446	5461	5476	5490	5505	5519	5534	5548	5563	5577	2	5	7	10	12
34	·5592	5606	5621	5635	5650	5664	5678	5693	5707	5721	2	5	7	10	12
35	·5736	5750	5764	5779	5793	5807	5821	5835	5850	5864	2	5	7	10	12
36	·5878	5892	5906	5920	5934	5948	5962	5976	5990	6004	2	5	7	9	12
37	·6018	6032	6046	6060	6074	6088	6101	6115	6129	6143	2	5	7	9	12
38	·6157	6170	6184	6198	6211	6225	6239	6252	6266	6280	2	5	7	9	11
39	·6293	6307	6320	6334	6347	6361	6374	6388	6401	6414	2	4	7	9	11
40	·6428	6441	6455	6468	6481	6494	6508	6521	6534	6547	2	4	7	9	11
41	·6561	6574	6587	6600	6613	6626	6639	6652	6665	6678	2	4	7	9	11
42	·6691	6704	6717	6730	6743	6756	6769	6782	6794	6807	2	4	6	9	11
43	·6820	6833	6845	6858	6871	6884	6896	6909	6921	6934	2	4	6	8	11
44	·6947	6959	6972	6984	6997	7009	7022	7034	7046	7059	2	4	6	8	10

NATURAL SINES

°	0′	6′	12′	18′	24′	30′	36′	42′	48′	54′	1′	2′	3′	4′	5′
45	·7071	7083	7096	7108	7120	7133	7145	7157	7169	7181	2	4	6	8	10
46	·7193	7206	7218	7230	7242	7254	7266	7278	7290	7302	2	4	6	8	10
47	·7314	7325	7337	7349	7361	7373	7385	7396	7408	7420	2	4	6	8	10
48	·7431	7443	7455	7466	7478	7490	7501	7513	7524	7536	2	4	6	8	10
49	·7547	7559	7570	7581	7593	7604	7615	7627	7638	7649	2	4	6	8	9
50	·7660	7672	7683	7694	7705	7716	7727	7738	7749	7760	2	4	6	7	9
51	·7771	7782	7793	7804	7815	7826	7837	7848	7859	7869	2	4	5	7	9
52	·7880	7891	7902	7912	7923	7934	7944	7955	7965	7976	2	4	5	7	9
53	·7986	7997	8007	8018	8028	8039	8049	8059	8070	8080	2	3	5	7	9
54	·8090	8100	8111	8121	8131	8141	8151	8161	8171	8181	2	3	5	7	8
55	·8192	8202	8211	8221	8231	8241	8251	8261	8271	8281	2	3	5	7	8
56	·8290	8300	8310	8320	8329	8339	8348	8358	8368	8377	2	3	5	6	8
57	·8387	8396	8406	8415	8425	8434	8443	8453	8462	8471	2	3	5	6	8
58	·8480	8490	8499	8508	8517	8526	8536	8545	8554	8563	2	3	5	6	8
59	·8572	8581	8590	8599	8607	8616	8625	8634	8643	8652	1	3	4	6	7
60	·8660	8669	8678	8686	8695	8704	8712	8721	8729	8738	1	3	4	6	7
61	·8746	8755	8763	8771	8780	8788	8796	8805	8813	8821	1	3	4	6	7
62	·8829	8838	8846	8854	8862	8870	8878	8886	8894	8902	1	3	4	5	7
63	·8910	8918	8926	8934	8942	8949	8957	8965	8973	8980	1	3	4	5	6
64	·8988	8996	9003	9011	9018	9026	9033	9041	9048	9056	1	3	4	5	6
65	·9063	9070	9078	9085	9092	9100	9107	9114	9121	9128	1	2	4	5	6
66	·9135	9143	9150	9157	9164	9171	9178	9184	9191	9198	1	2	3	5	6
67	·9205	9212	9219	9225	9232	9239	9245	9252	9259	9265	1	2	3	4	6
68	·9272	9278	9285	9291	9298	9304	9311	9317	9323	9330	1	2	3	4	5
69	·9336	9342	9348	9354	9361	9367	9373	9379	9385	9391	1	2	3	4	5
70	·9397	9403	9409	9415	9421	9426	9432	9438	9444	9449	1	2	3	4	5
71	·9455	9461	9466	9472	9478	9483	9489	9494	9500	9505	1	2	3	4	5
72	·9511	9516	9521	9527	9532	9537	9542	9548	9553	9558	1	2	3	4	4
73	·9563	9568	9573	9578	9583	9588	9593	9598	9603	9608	1	2	2	3	4
74	·9613	9617	9622	9627	9632	9636	9641	9646	9650	9655	1	2	2	3	4
75	·9659	9664	9668	9673	9677	9681	9686	9690	9694	9699	1	1	2	3	4
76	·9703	9707	9711	9715	9720	9724	9728	9732	9736	9740	1	1	2	3	3
77	·9744	9748	9751	9755	9759	9763	9767	9770	9774	9778	1	1	2	3	3
78	·9781	9785	9789	9792	9796	9799	9803	9806	9810	9813	1	1	2	2	3
79	·9816	9820	9823	9826	9829	9833	9836	9839	9842	9845	1	1	2	2	3
80	·9848	9851	9854	9857	9860	9863	9866	9869	9871	9874	0	1	1	2	2
81	·9877	9880	9882	9885	9888	9890	9893	9895	9898	9900	0	1	1	2	2
82	·9903	9905	9907	9910	9912	9914	9917	9919	9921	9923	0	1	1	2	2
83	·9925	9928	9930	9932	9934	9936	9938	9940	9942	9943	0	1	1	1	2
84	·9945	9947	9949	9951	9952	9954	9956	9957	9959	9960	0	1	1	1	1
85	·9962	9963	9965	9966	9968	9969	9971	9972	9973	9974	0	0	1	1	1
86	·9976	9977	9978	9979	9980	9981	9982	9983	9984	9985	0	0	1	1	1
87	·9986	9987	9988	9989	9990	9990	9991	9992	9993	9993	0	0	0	0	0
88	·9994	9995	9995	9996	9996	9997	9997	9997	9998	9998	0	0	0	0	0
89	·9998	9999	9999	9999	9999	1·000	1·000	1·000	1·000	1·000	0	0	0	0	0

NATURAL COSINES

°	0′	6′	12′	18′	24′	30′	36′	42′	48′	54′	SUBTRACT 1′	2′	3′	4′	5′
0	1·000	1·000	1·000	1·000	1·000	1·000	**·9999**	**9999**	**9999**	**9999**	0	0	0	0	0
1	·9998	9998	9998	9997	9997	9997	9996	9996	9995	9995	0	0	0	0	0
2	·9994	9993	9993	9992	9991	9990	9990	9989	9988	9987	0	0	0	0	0
3	·9986	9985	9984	9983	9982	9981	9980	9979	9978	9977	0	0	1	1	1
4	·9976	9974	9973	9972	9971	9969	9968	9966	9965	9963	0	0	1	1	1
5	·9962	9960	9959	9957	9956	9954	9952	9951	9949	9947	0	1	1	1	1
6	·9945	9943	9942	9940	9938	9936	9934	9932	9930	9928	0	1	1	1	2
7	·9925	9923	9921	9919	9917	9914	9912	9910	9907	9905	0	1	1	2	2
8	·9903	9900	9898	9895	9893	9890	9888	9885	9882	9880	0	1	1	2	2
9	·9877	9874	9871	9869	9866	9863	9860	9857	9854	9851	0	1	1	2	2
10	·9848	9845	9842	9839	9836	9833	9829	9826	9823	9820	1	1	2	2	3
11	·9816	9813	9810	9806	9803	9799	9796	9792	9789	9785	1	1	2	2	3
12	·9781	9778	9774	9770	9767	9763	9759	9755	9751	9748	1	1	2	3	3
13	·9744	9740	9736	9732	9728	9724	9720	9715	9711	9707	1	1	2	3	3
14	·9703	9699	9694	9690	9686	9681	9677	9673	9668	9664	1	1	2	3	4
15	·9659	9655	9650	9646	9641	9636	9632	9627	9622	9617	1	2	2	3	4
16	·9613	9608	9603	9598	9593	9588	9583	9578	9573	9568	1	2	2	3	4
17	·9563	9558	9553	9548	9542	9537	9532	9527	9521	9516	1	2	3	4	4
18	·9511	9505	9500	9494	9489	9483	9478	9472	9466	9461	1	2	3	4	5
19	·9455	9449	9444	9438	9432	9426	9421	9415	9409	9403	1	2	3	4	5
20	·9397	9391	9385	9379	9373	9367	9361	9354	9348	9342	1	2	3	4	5
21	·9336	9330	9323	9317	9311	9304	9298	9291	9285	9278	1	2	3	4	5
22	·9272	9265	9259	9252	9245	9239	9232	9225	9219	9212	1	2	3	4	6
23	·9205	9198	9191	9184	9178	9171	9164	9157	9150	9143	1	2	3	5	6
24	·9135	9128	9121	9114	9107	9100	9092	9085	9078	9070	1	2	4	5	6
25	·9063	9056	9048	9041	9033	9026	9018	9011	9003	8996	1	3	4	5	6
26	·8988	8980	8973	8965	8957	8949	8942	8934	8926	8918	1	3	4	5	6
27	·8910	8902	8894	8886	8878	8870	8862	8854	8846	8838	1	3	4	5	7
28	·8829	8821	8813	8805	8796	8788	8780	8771	8763	8755	1	3	4	6	7
29	·8746	8738	8729	8721	8712	8704	8695	8686	8678	8669	1	3	4	6	7
30	·8660	8652	8643	8634	8625	8616	8607	8599	8590	8581	1	3	4	6	7
31	·8572	8563	8554	8545	8536	8526	8517	8508	8499	8490	2	3	5	6	8
32	·8480	8471	8462	8453	8443	8434	8425	8415	8406	8396	2	3	5	6	8
33	·8387	8377	8368	8358	8348	8339	8329	8320	8310	8300	2	3	5	6	8
34	·8290	8281	8271	8261	8251	8241	8231	8221	8211	8202	2	3	5	7	8
35	·8192	8181	8171	8161	8151	8141	8131	8121	8111	8100	2	3	5	7	8
36	·8090	8080	8070	8059	8049	8039	8028	8018	8007	7997	2	3	5	7	9
37	·7986	7976	7965	7955	7944	7934	7923	7912	7902	7891	2	4	5	7	9
38	·7880	7869	7859	7848	7837	7826	7815	7804	7793	7782	2	4	5	7	9
39	·7771	7760	7749	7738	7727	7716	7705	7694	7683	7672	2	4	6	7	9
40	·7660	7649	7638	7627	7615	7604	7593	7581	7570	7559	2	4	6	8	9
41	·7547	7536	7524	7513	7501	7490	7478	7466	7455	7443	2	4	6	8	10
42	·7431	7420	7408	7396	7385	7373	7361	7349	7337	7325	2	4	6	8	10
43	·7314	7302	7290	7278	7266	7254	7242	7230	7218	7206	2	4	6	8	10
44	·7193	7181	7169	7157	7145	7133	7120	7108	7096	7083	2	4	6	8	10

Figures in **bold** type show change of integer

SUBTRACT

NATURAL COSINES

											SUBTRACT				
°	0′	6′	12′	18′	24′	30′	36′	42′	48′	54′	1′	2′	3′	4′	5′
45	·7071	7059	7046	7034	7022	7009	6997	6984	6972	6959	2	4	6	8	10
46	·6947	6934	6921	6909	6896	6884	6871	6858	6845	6833	2	4	6	8	11
47	·6820	6807	6794	6782	6769	6756	6743	6730	6717	6704	2	4	6	9	11
48	·6691	6678	6665	6652	6639	6626	6613	6600	6587	6574	2	4	7	9	11
49	·6561	6547	6534	6521	6508	6494	6481	6468	6455	6441	2	4	7	9	11
50	·6428	6414	6401	6388	6374	6361	6347	6334	6320	6307	2	4	7	9	11
51	·6293	6280	6266	6252	6239	6225	6211	6198	6184	6170	2	5	7	9	11
52	·6157	6143	6129	6115	6101	6088	6074	6060	6046	6032	2	5	7	9	12
53	·6018	6004	5990	5976	5962	5948	5934	5920	5906	5892	2	5	7	9	12
54	·5878	5864	5850	5835	5821	5807	5793	5779	5764	5750	2	5	7	9	12
55	·5736	5721	5707	5693	5678	5664	5650	5635	5621	5606	2	5	7	10	12
56	·5592	5577	5563	5548	5534	5519	5505	5490	5476	5461	2	5	7	10	12
57	·5446	5432	5417	5402	5388	5373	5358	5344	5329	5314	2	5	7	10	12
58	·5299	5284	5270	5255	5240	5225	5210	5195	5180	5165	2	5	7	10	12
59	·5150	5135	5120	5105	5090	5075	5060	5045	5030	5015	3	5	8	10	13
60	·5000	4985	4970	4955	4939	4924	4909	4894	4879	4863	3	5	8	10	13
61	·4848	4833	4818	4802	4787	4772	4756	4741	4726	4710	3	5	8	10	13
62	·4695	4679	4664	4648	4633	4617	4602	4586	4571	4555	3	5	8	10	13
63	·4540	4524	4509	4493	4478	4462	4446	4431	4415	4399	3	5	8	10	13
64	·4384	4368	4352	4337	4321	4305	4289	4274	4258	4242	3	5	8	11	13
65	·4226	4210	4195	4179	4163	4147	4131	4115	4099	4083	3	5	8	11	13
66	·4067	4051	4035	4019	4003	3987	3971	3955	3939	3923	3	5	8	11	13
67	·3907	3891	3875	3859	3843	3827	3811	3795	3778	3762	3	5	8	11	13
68	·3746	3730	3714	3697	3681	3665	3649	3633	3616	3600	3	5	8	11	14
69	·3584	3567	3551	3535	3518	3502	3486	3469	3453	3437	3	5	8	11	14
70	·3420	3404	3387	3371	3355	3338	3322	3305	3289	3272	3	5	8	11	14
71	·3256	3239	3223	3206	3190	3173	3156	3140	3123	3107	3	6	8	11	14
72	·3090	3074	3057	3040	3024	3007	2990	2974	2957	2940	3	6	8	11	14
73	·2924	2907	2890	2874	2857	2840	2823	2807	2790	2773	3	6	8	11	14
74	·2756	2740	2723	2706	2689	2672	2656	2639	2622	2605	3	6	8	11	14
75	·2588	2571	2554	2538	2521	2504	2487	2470	2453	2436	3	6	8	11	14
76	·2419	2402	2385	2368	2351	2334	2317	2300	2284	2267	3	6	8	11	14
77	·2250	2233	2215	2198	2181	2164	2147	2130	2113	2096	3	6	9	11	14
78	·2079	2062	2045	2028	2011	1994	1977	1959	1942	1925	3	6	9	11	14
79	·1908	1891	1874	1857	1840	1822	1805	1788	1771	1754	3	6	9	11	14
80	·1736	1719	1702	1685	1668	1650	1633	1616	1599	1582	3	6	9	11	14
81	·1564	1547	1530	1513	1495	1478	1461	1444	1426	1409	3	6	9	12	14
82	·1392	1374	1357	1340	1323	1305	1288	1271	1253	1236	3	6	9	12	14
83	·1219	1201	1184	1167	1149	1132	1115	1097	1080	1063	3	6	9	12	14
84	·1045	1028	1011	0993	0976	0958	0941	0924	0906	0889	3	6	9	12	14
85	·0872	0854	0837	0819	0802	0785	0767	0750	0732	0715	3	6	9	12	14
86	·0698	0680	0663	0645	0628	0610	0593	0576	0558	0541	3	6	9	12	15
87	·0523	0506	0488	0471	0454	0436	0419	0401	0384	0366	3	6	9	12	15
88	·0349	0332	0314	0297	0279	0262	0244	0227	0209	0192	3	6	9	12	15
89	·0175	0157	0140	0122	0105	0087	0070	0052	0035	0017	3	6	9	12	15

SUBTRACT

NATURAL TANGENTS

°	0′	6′	12′	18′	24′	30′	36′	42′	48′	54′	1′	2′	3′	4′	5′
0	·0000	0017	0035	0052	0070	0087	0105	0122	0140	0157	3	6	9	12	15
1	·0175	0192	0209	0227	0244	0262	0279	0297	0314	0332	3	6	9	12	15
2	·0349	0367	0384	0402	0419	0437	0454	0472	0489	0507	3	6	9	12	15
3	·0524	0542	0559	0577	0594	0612	0629	0647	0664	0682	3	6	9	12	15
4	·0699	0717	0734	0752	0769	0787	0805	0822	0840	0857	3	6	9	12	15
5	·0875	0892	0910	0928	0945	0963	0981	0998	1016	1033	3	6	9	12	15
6	·1051	1069	1086	1104	1122	1139	1157	1175	1192	1210	3	6	9	12	15
7	·1228	1246	1263	1281	1299	1317	1334	1352	1370	1388	3	6	9	12	15
8	·1405	1423	1441	1459	1477	1495	1512	1530	1548	1566	3	6	9	12	15
9	·1584	1602	1620	1638	1655	1673	1691	1709	1727	1745	3	6	9	12	15
10	·1763	1781	1799	1817	1835	1853	1871	1890	1908	1926	3	6	9	12	15
11	·1944	1962	1980	1998	2016	2035	2053	2071	2089	2107	3	6	9	12	15
12	·2126	2144	2162	2180	2199	2217	2235	2254	2272	2290	3	6	9	12	15
13	·2309	2327	2345	2364	2382	2401	2419	2438	2456	2475	3	6	9	12	15
14	·2493	2512	2530	2549	2568	2586	2605	2623	2642	2661	3	6	9	12	16
15	·2679	2698	2717	2736	2754	2773	2792	2811	2830	2849	3	6	9	13	16
16	·2867	2886	2905	2924	2943	2962	2981	3000	3019	3038	3	6	9	13	16
17	·3057	3076	3096	3115	3134	3153	3172	3191	3211	3230	3	6	10	13	16
18	·3249	3269	3288	3307	3327	3346	3365	3385	3404	3424	3	6	10	13	16
19	·3443	3463	3482	3502	3522	3541	3561	3581	3600	3620	3	7	10	13	16
20	·3640	3659	3679	3699	3719	3739	3759	3779	3799	3819	3	7	10	13	17
21	·3839	3859	3879	3899	3919	3939	3959	3979	4000	4020	3	7	10	13	17
22	·4040	4061	4081	4101	4122	4142	4163	4183	4204	4224	3	7	10	14	17
23	·4245	4265	4286	4307	4327	4348	4369	4390	4411	4431	3	7	10	14	17
24	·4452	4473	4494	4515	4536	4557	4578	4599	4621	4642	4	7	11	14	18
25	·4663	4684	4706	4727	4748	4770	4791	4813	4834	4856	4	7	11	14	18
26	·4877	4899	4921	4942	4964	4986	5008	5029	5051	5073	4	7	11	15	18
27	·5095	5117	5139	5161	5184	5206	5228	5250	5272	5295	4	7	11	15	18
28	·5317	5340	5362	5384	5407	5430	5452	5475	5498	5520	4	8	11	15	19
29	·5543	5566	5589	5612	5635	5658	5681	5704	5727	5750	4	8	12	15	19
30	·5774	5797	5820	5844	5867	5890	5914	5938	5961	5985	4	8	12	16	20
31	·6009	6032	6056	6080	6104	6128	6152	6176	6200	6224	4	8	12	16	20
32	·6249	6273	6297	6322	6346	6371	6395	6420	6445	6469	4	8	12	16	20
33	·6494	6519	6544	6569	6594	6619	6644	6669	6694	6720	4	8	13	17	21
34	·6745	6771	6796	6822	6847	6873	6899	6924	6950	6976	4	9	13	17	21
35	·7002	7028	7054	7080	7107	7133	7159	7186	7212	7239	4	9	13	18	22
36	·7265	7292	7319	7346	7373	7400	7427	7454	7481	7508	5	9	14	18	23
37	·7536	7563	7590	7618	7646	7673	7701	7729	7757	7785	5	9	14	18	23
38	·7813	7841	7869	7898	7926	7954	7983	8012	8040	8069	5	9	14	19	24
39	·8098	8127	8156	8185	8214	8243	8273	8302	8332	8361	5	10	15	20	24
40	·8391	8421	8451	8481	8511	8541	8571	8601	8632	8662	5	10	15	20	25
41	·8693	8724	8754	8785	8816	8847	8878	8910	8941	8972	5	10	16	21	26
42	·9004	9036	9067	9099	9131	9163	9195	9228	9260	9293	5	11	16	21	27
43	·9325	9358	9391	9424	9457	9490	9523	9556	9590	9623	6	11	17	22	28
44	·9657	9691	9725	9759	9793	9827	9861	9896	9930	9965	6	11	17	23	29

NATURAL TANGENTS

°	0′	6′	12′	18′	24′	30′	36′	42′	48′	54′	1′	2′	3′	4′	5′
45	1·0000	0035	0070	0105	0141	0176	0212	0247	0283	0319	6	12	18	24	30
46	1·0355	0392	0428	0464	0501	0538	0575	0612	0649	0686	6	12	18	25	31
47	1·0724	0761	0799	0837	0875	0913	0951	0990	1028	1067	6	13	19	25	32
48	1·1106	1145	1184	1224	1263	1303	1343	1383	1423	1463	7	13	20	26	33
49	1·1504	1544	1585	1626	1667	1708	1750	1792	1833	1875	7	14	21	28	34
50	1·1918	1960	2002	2045	2088	2131	2174	2218	2261	2305	7	14	22	29	36
51	1·2349	2393	2437	2482	2527	2572	2617	2662	2708	2753	8	15	23	30	38
52	1·2799	2846	2892	2938	2985	3032	3079	3127	3175	3222	8	16	24	31	39
53	1·3270	3319	3367	3416	3465	3514	3564	3613	3663	3713	8	16	25	33	41
54	1·3764	3814	3865	3916	3968	4019	4071	4124	4176	4229	9	17	26	34	43
55	1·4281	4335	4388	4442	4496	4550	4605	4659	4715	4770	9	18	27	36	45
56	1·4826	4882	4938	4994	5051	5108	5166	5224	5282	5340	10	19	29	38	48
57	1·5399	5458	5517	5577	5637	5697	5757	5818	5880	5941	10	20	30	40	50
58	1·6003	6066	6128	6191	6255	6319	6383	6447	6512	6577	11	21	32	43	53
59	1·6643	6709	6775	6842	6909	6977	7045	7113	7182	7251	11	23	34	45	56
60	1·7321	7391	7461	7532	7603	7675	7747	7820	7893	7966	12	24	36	48	60
61	1·8040	8115	8190	8265	8341	8418	8495	8572	8650	8728	13	26	38	51	64
62	1·8807	8887	8967	9047	9128	9210	9292	9375	9458	9542	14	27	41	55	68
63	1 9626	9711	9797	9883	9970	**0057**	**0145**	**0233**	**0323**	**0413**	15	29	44	58	73
64	2·0503	0594	0686	0778	0872	0965	1060	1155	1251	1348	16	31	47	63	78
65	2·1445	1543	1642	1742	1842	1943	2045	2148	2251	2355	17	34	51	68	85
66	2·2460	2566	2673	2781	2889	2998	3109	3220	3332	3445	18	37	55	73	92
67	2·3559	3673	3789	3906	4023	4142	4262	4383	4504	4627	20	40	60	79	99
68	2·4751	4876	5002	5129	5257	5386	5517	5649	5782	5916	22	43	65	87	108
69	2·6051	6187	6325	6464	6605	6746	6889	7034	7179	7326	24	47	71	95	118
70	2·7475	7625	7776	7929	8083	8239	8397	8556	8716	8878	26	52	78	104	130
71	2·9042	9208	9375	9544	9714	9887	**0061**	**0237**	**0415**	**0595**	29	58	87	116	145
72	3·0777	0961	1146	1334	1524	1716	1910	2106	2305	2506	32	64	96	129	161
73	3·2709	2914	3122	3332	3544	3759	3977	4197	4420	4646	36	72	108	144	180
74	3·4874	5105	5339	5576	5816	6059	6305	6554	6806	7062	41	81	122	163	203
75	3·7321	7583	7848	8118	8391	8667	8947	9232	9520	9812	**Use interpolation**				
76	4·0108	0408	0713	1022	1335	1653	1976	2303	2635	2972					
77	4·3315	3662	4015	4373	4737	5107	5483	5864	6252	6646					
78	4·7046	7453	7867	8288	8716	9152	9594	**0045**	**0504**	**0970**					
79	5·1446	1929	2422	2924	3435	3955	4486	5026	5578	6140					
80	5·6713	7297	7894	8502	9124	9758	**0405**	**1066**	**1742**	**2432**					
81	6·3138	3859	4596	5350	6122	6912	7720	8548	9395	**0264**					
82	7·1154	2066	3002	3962	4947	5958	6996	8062	9158	**0285**					
83	8·1443	2636	3863	5126	6427	7769	9152	**0579**	**2052**	**3572**					
84	9·5144	6768	8448	10·02	10·20	10·39	10·58	10·78	10·99	11·20					
85	11·43	11·66	11·91	12·16	12·43	12·71	13·00	13·30	13·62	13·95					
86	14·30	14·67	15·06	15·46	15·89	16·35	16·83	17·34	17·89	18·46					
87	19·08	19·74	20·45	21·20	22·02	22·90	23·86	24·90	26·03	27·27					
88	28·64	30·14	31·82	33·69	35·80	38·19	40·92	44·07	47·74	52·08					
89	57·29	63·66	71·62	81·85	95·49	114·6	143·2	191·0	286·5	573·0					

Figures in **bold type** show changes of integer

Appendix 5

Statistical appendix

A.5.1 Standard deviation, mean, mode and median

Taking any sample population of data ($x_1, x_2, \ldots, x_n$) the *arithmetic mean* is given by:

$$\bar{x} = \sum_{i=1}^{i=n} x_i$$

It is as well to note that only very rarely are we able to calculate the *true* mean (μ) for a total population and we must expect our sample arithmetic mean to differ somewhat from the true mean. The *mode* of a sample is the value or value class (e.g. 1·0–1·9) which occurs most frequently in the sample. The modal value is the crest of the frequency distribution curve (see section 7.4). Some distributions exhibit bimodal (twin peak) distributions, or even a greater number of peaks. The *median* value is the central value in the string $x_1, x_2, \ldots, x_n$ when they are ranked in order. As an illustration let us take a sample of 21 values:

7, 5, 3, 6, 9, 4, 2, 4, 11, 13, 5, 4, 3, 3, 4, 6, 4, 8, 10, 12, 1

The mean is 5·91 and ranking the values in ascending order we have:

1, 2, 3, 3, 3, 4, 4, 4, 4, 4, **5**, 5, 6, 6, 7, 8, 9, 10, 11, 12, 13,

whilst the most frequently occurring value, the mode, is 4. The frequency distribution is shown in figure A.1. Here we see that the mode lies to the left of the mean value, and we term the distribution *positively skewed*, with a relatively large number of values below the mean value. If the modal value had lain to the right of the mean value, then the distribution would be *negatively skewed.*

One particular property of the mean is that if we sum the *deviations* of values from the mean $(\bar{x}-x_1)+(\bar{x}-x_2)+\ldots+(\bar{x}-x_n)$ then we arrive at zero. If we were to square each of the deviations from the mean and take the sum of the squares of the deviations, then this sum would be less than the sum of the squares of deviations from any other point. This is an important property and we term the sum of the squares of the deviations from the mean, divided by the sample size (n), the *variance* (σ^2). This is a measure of the dispersion of data values about the mean point. In the example above the variance is 13·7 indicating a relatively wide dispersion of values about the mean as we saw in figure A.1. If we were to take the sequence 3, 4, 4, 4, 3 as our sample, then the mean would be 3·6 and the variance only 0·24, and for 3, 3, 3, 3, 3, a mean of 3·0 and a variance of 0·0.

The *standard deviation* (σ) is the square root of the variance so that:

$$\sigma = \sqrt{\frac{1}{n}\sum_{i=1}^{i=n} (\bar{x}-x_i)^2}$$

A.5.2. Correlation coefficients

The correlation coefficient expresses the degree to which variation in one variable is associated with variation in another. It is effectively a measure of the

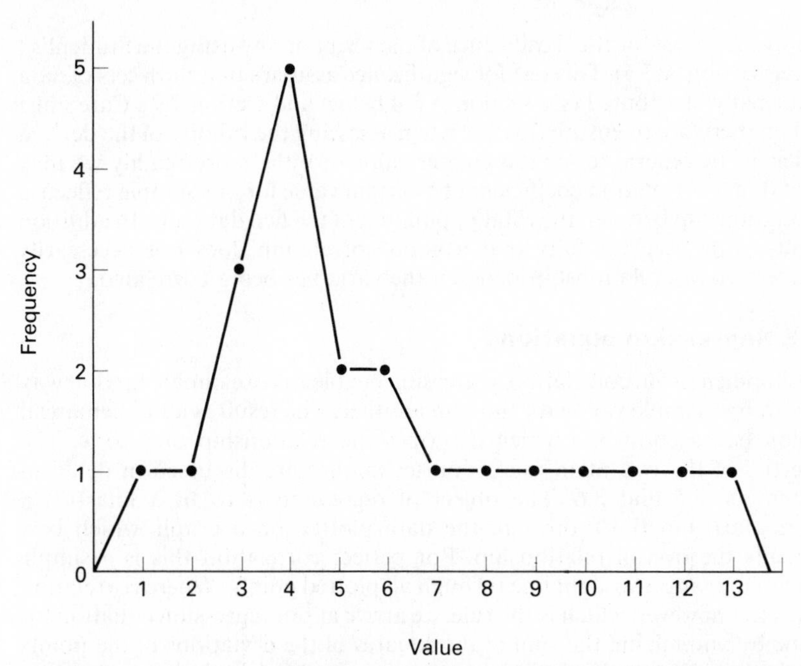

covariance of the two populations. For a perfect correlation between two variable sets where y increases exactly as x increases we have a correlation coefficient of 1·0. For a perfect correlation when y increases exactly as x *decreases* the correlation coefficient is $-1{\cdot}0$. In these cases if we were to plot all data points of x and y on a graph the plot of points will lie in a straight line (see regression, section A.5.3 below) with positive or negative slope respectively (see section 3.1). As the covariance of x and y becomes less, the degree of correlation decreases and the scatter of points over the graph becomes greater so that we have total non-correlation where the correlation coefficient is zero.

Two correlation coefficients are used in statistical analysis of covariation between pairs of data sets. The first of these uses data in ranked form, and is given by:

$$r = 1 - \frac{6 \sum_{i=1}^{i=n} d_i^2}{n^3 - n}$$

where r is known as the Spearman's rank correlation coefficient, and where

$$d_i = (\bar{x} - x_i) - (\bar{y} - y_i) = y_i - x_i$$

since in ranks $\bar{x} = \bar{y}$.

This measure of correlation is not as powerful statistically as the product-moment correlation coefficient below, and it is far less useful. The product-moment correlation coefficient is defined as:

$$r = \frac{\sum_{i=1}^{i=n} (\bar{x} - x_i)(\bar{y} - y_i)}{\sqrt{\sigma_x^2 \sigma_y^2}} \quad \text{(covariance of } x \text{ and } y\text{)}$$

We are able to test for the significance of the result of r by using the Student's t test (see section A.5.5). This test for significance assumes that both sets of data are normally distributed (see section A.5.4 below and section 7.7). Care must be taken therefore to ensure this fact when assessing the validity of the derived coefficient. In general terms the greater value of n the more readily we may accept that a correlation coefficient of a certain value for our sample reflects a true relationship between the total population of the two data sets. In addition we must note very carefully that a good correlation does not necessarily indicate a *causal* relationship between the variables being correlated.

A.5.3 Regression equations

A development from correlation, regression enables us to estimate in what way values in one sample vary with those in another. The result is a mathematical function (see section 3.1) which describes the relationship of y to x. The properties of the two main regression techniques are discussed in detail in sections 3.4, 3.5 and 3.6. The object of regression is to fit a line (i.e. a mathematical function), through the data plotted on a graph, which best represents the overall relationship. For perfect correlation this is a simple matter of drawing a straight line through all plotted points. Where correlation is imperfect however, which is the rule, we arrive at our regression equation for the line by minimizing the sum of the squares of the deviations of the points from the line. Where the deviations are minimized resolved along the y-axis (see figure 3.7) we have the regression of y on x (the normal case assuming x to be the independent variable) and the equation is:

$$y - \bar{y} = r\frac{\sigma_y}{\sigma_x}(x - \bar{x})$$

and the regression of y on x is:

$$x - \bar{x} = r\frac{\sigma_x}{\sigma_y}(y - \bar{y})$$

There is an unique association between the value of the correlation coefficient and relationship between the two regression lines, where the cosine of the angle between the two lines (θ) is given by r, so that $r = \cos\theta$. Thus when we have perfect correlation the two lines are coincident, and where correlation is zero the two lines are at right-angles. This means of obtaining regression lines is known as the process of *least-squares regression.* Again, if we are to use regression equations for predictive purposes we shall need to ensure that data are normally distributed.

The second method for obtaining a regression-line equation does not use the correlation coefficient and minimizes the deviations of both x and y points from the line. This is the *reduced major axis* regression line and is given by:

$$y = \frac{\sigma_y}{\sigma_x}x + k$$

where k is determined by the substitution of $\bar{x}$ and $\bar{y}$ into the equation. This line and the two least-squares regression lines pass through the means of both sample populations.

A.5.4 The normal distribution curve

The mathematics of the normal curve are dealt with in section 7.7 and its statistical meaning and usage are given here. Data falls into three major categories based upon the *level of measurement*. These are:

Nominal: where data exist only in classes, for example, for soil type, podzol or brown earth, and where the choice of available statistics is limited by the nature of the original data, and the statistical result is of limited usefulness.
Ordinal: where data are *ranked* for the purpose of carrying out statistical tests. The statistics used are of greater usefulness and their results have greater power than those used for nominal data.
Interval: where actual numeric values are used in the statistics. The most powerful statistical tests use interval data.

Statistical methods largely fall into two categories, those using nominal or ordinal data, called non-parametric statistics, and those using interval data, called parametric statistics. Parametric statistics generally assume that data are approximately normally distributed.

The normal curve has certain unique statistical properties, particularly in terms of the standard deviation, making it useful in assessing probabilities of occurrence and the significance of results. For the normal curve, 68·26% of the area under the curve lies between $\bar{x}+\sigma$ and $\bar{x}-\sigma$. Thus 68·26% of a normally distributed sample will have values lying one standard deviation on either side of the mean. Similarly 95·44% lie between $\bar{x}+2\sigma$ and $\bar{x}-2\sigma$, and 99·73% lie between $\bar{x}+3\sigma$ and $\bar{x}-3\sigma$. The normal curve extends asymptotically in each direction and so in theory no finite number of standard deviations will include all the total area under the curve, but so little will lie beyond $\bar{x}+4\sigma$ and $\bar{x}-4\sigma$ that we can often ignore these extremes of the curve.

Often data do not approximate the normal distribution closely enough to warrant them being considered normally distributed for statistical purposes. When this is so we can sometimes transform the data by applying an appropriate mathematical function according to whether the data are positively or negatively skewed. Generally the data are transformed using the logarithm of data values for a positive skewness, and squaring the data for negative skewness.

A.5.5 Standard error of estimate of y

Statistics are necessary because more often than not we are aiming to draw conclusions about a phenomonon or relationship based only upon a relatively small sample of the total population. As a result of this we need some means of judging how far our result obtained from the sample will differ from the true result for the entire population.

For any infinite total population we are able to estimate the true mean (μ) by calculating a sample mean ($\bar{x}$), and to estimate the true standard deviation (σ_x) by calculating the sample standard deviation (s). We can repeat the process of estimation a very large number of times, each time drawing a different sample from the total population, and each time attaining different reults. For very small samples there is a relatively greater chance of one or two very unusual (very high or very low) values drastically altering the value of the sample mean and standard deviation. Any measure of the probable difference of the sample mean from the true mean will thus include n. In addition the chances of calculating a highly spurious sample mean are dramatically increased as the dispersion (i.e. the standard deviation) of the data values about the mean is increased. We thus use the *standard error or the mean* ($s_{\bar{x}}^2$), where:

$$s_{\bar{x}} = s/\sqrt{n}$$

as an index of the difference between sample and true means. In fact the standard error of the mean is the standard deviation of sample means about the true mean. This can be developed to provide an estimator of the validity of predictions made using regression. Here, the *standard error of the estimate* of y from x is given by':

$$s_{y \cdot x} = \sigma_y \sqrt{(1 - r^2)}$$

We are assuming data to be normally distributed, and as long as both sets are so distributed we can say that 95·44% of estimated values of y from the regression will lie between $\pm 2\ s_{y \cdot x}$ and 99·73% will lie between $\pm 3\ s_{y \cdot x}$. It is therefore customary to insert 'confidence limits' on either side of a regression line on a graph, generally at $\pm 2\ s_{y \cdot x}$, indicating the range of approximately 95% of the values of y—the 95% confidence limits.

A.5.6 Student's *t*–test

The student's t-test is used to test the hypothesis that a sample of mean $\bar{x}$ could have been drawn from a total population whose mean is μ. It is thus a development of the standard error of the mean, and is given by:

$$t = \frac{\text{difference between sample and population means}}{\text{standard error of the mean}}$$

$$= \frac{\mu - \bar{x}}{\sigma/\sqrt{n}}$$

The value of t obtained by this formula must then be compared with calculated values of t for different probabilities. These are given in most statistical texts. Again the significance of any one result is increased if we used a larger sample size. In general, for constant n, a higher t-value indicates that there is only a small probability that the sample was *not* drawn from the total population.

A.5.7 Multiple correlation coefficient

The multiple correlation coefficient expresses the degree of linear correlation between a number of variables (say, $x_1, x_2, \ldots, x_n$). The coefficient for a three-variable case is $R_{1 \cdot 23}$, where the first subscript is the dependent, and the final pair are the independent variables, and can be usefully related to simple correlation coefficients for different pairs of data sets. Where r_{12} is the simple correlation coefficient between set 1 and set 2, r_{23} between set 2 and set 3, then for the three variable case:

$$R_{1.23} = \sqrt{\frac{r_{13}^2 + r_{12}^2 - 2r_{13}r_{12}r_{32}}{1 - r_{32}^2}}$$

The resultant coefficient always takes the value $0 \leq R \leq 1$, and R^2 is the variance in x_1 accounted for by x_2 and x_3.

Suggestions for further reading

Chapter 1

On general mathematical concepts and conventions

COURANT, R. and ROBBINS, H. 1973: *What is mathematics*? London: Oxford University Press.

GOODSTEIN, R. L. 1962: *Fundamental concepts of mathematics*. London: Pergamon.

WILDER, R. L. 1968: *Evolution of mathematical concepts: an elementary study*. New York: Wiley.

On logic and set theory

BISHIR, J. W. and DREWES, D. W. 1970: *Mathematics in the behavioural and social sciences*. New York: Harcourt, Brace and World.

MIZRAHI, A. and SULLIVAN, M. 1973: *Finite mathematics with applications for business and social sciences*. New York: Wiley.

On units

NATIONAL PHYSICAL LABORATORY, 1970: *SI the international system of units*. London: HMSO.

Chapter 3

On systems

CHORLEY, R. J. and KENNEDY, B. A. 1971: *Physical geography: a systems approach*. London: Prentice Hall.

RUSSWURM, L. H. and SOMMERVILLE, E. 1974: *Man's natural environment: a systems approach*. North Scituate, Mass.: Duxbury.

On general statistical techniques

BRYANT, E. C. 1966: *Statistical analysis*. New York: McGraw-Hill.

GREGORY, S. 1970: *Statistical methods and the geographer*, second edition. London: Longman.

SIEGEL, S. 1956: *Non-parametric statistics*. New York: McGraw-Hill.

On regression analysis

DRAPER, N. and SMITH, H. 1966: *Applied regression analysis*. New York: Wiley.

KRUSKAL, W. 1953: On the uniqueness of the line of organic correlation. *Biometrics* 9, 47–58.

MILLER, R. L. and KAHN, J. S. 1962: *Statistical analysis in the geological sciences*. London: Wiley.

On multivariate statistics

EZEKIEL, M. and FOX, K. A. 1959: *Methods of correlation and regression analysis—linear and curvilinear*. New York: Wiley.

PANOFSKY, H. A. and BRIER, G. W. 1958: *Some applications of statistics to meteorology*. Pennsylvania: Pennsylvania State University.

Chapter 4

On principal components analysis

DAULTREY, S. 1976: Principal components analysis. *Concepts and techniques in modern geography*. Norwich: Geo Abstracts.

KENDALL, M. G. 1965: *A course in multivariate analysis. Griffin Statistics Monograph*. London: Griffin.

HOPE, K. 1968: *Methods of multivariate analysis*. London: University of London Press.

On matrices

BRANFIELD, J. R. and BELL, H. W. 1970: *Matrices and their applications. Introductory Monograph in Mathematics*. London: Macmillan.

WILLIAMS, I. P. 1972: *Matrices for scientists*. Hutchinson University Library. London: Hutchinson.

Chapter 5

On coordinate geometry

FLANDERS, H. and PRICE, J. J. 1973: *Introductory college mathematics with linear algebra and finite mathematics*. New York: Academic Press.

On differentiation

KNIGHT, B. and ADAMS, R. 1975: *Calculus I* and *Calculus II*. London: Unwin.

STEPHENSON, G. 1973: *Mathematical methods for science students*. London: Longmans.

On curve fitting and numerical analysis

EZEKIEL, M. and FOX, K. A. 1959: *Methods of correlation and regression analysis—linear and curvilinear*. New York: Wiley.

HILDEBRAND, F. B. 1974: *Introduction to numerical analysis*. New York: McGraw-Hill.

KHABAZA, I. M. 1965: *Numerical analysis*. London: Pergamon.

WILLIAMS, E. J. 1959: *Regression analysis*. New York: Wiley.

Chapter 6

BISHIR, J. W. and DREWES, D. W. 1970: *Mathematics in the behavioural and social sciences*. Harcourt, Brace and World.

LANG, S. 1973: *A first course in calculus*. Reading, Massachusetts: Addison-Wesley.

Chapter 7

On probability and scientific method

DAVIDSON, D. A. 1978: *Science for physical geographers*. London: Edward Arnold.

On general probability

DURRAN, J. H. 1970: *SMP Statistics and probability*. Cambridge: Cambridge University Press.

HEATHCOTE, C. P. 1971: *Probability: elements of the mathematical theory. Unwin University Books*. London: Unwin.

On probability applications in hydrology and distributions

CHOW, V. T. 1964: *Handbook of applied hydrology*. New York: McGraw-Hill.

On Pearson distributions

DAWSON, J. A. 1972: The Pearson system of frequency curves. *Computer Applications in the natural and social Sciences* 13A, 1–10.

ELDERTON, W. P. and JOHNSON, N. L. 1969: *Systems of frequency curves*. Cambridge: Cambridge University Press.

On extremal and lognormal distributions

AITCHISON, J. and BROWN, J. A. C. 1957: *The lognormal distribution, with special reference to its uses in economics*. University of Cambridge Department of Applied Economics Monograph 5. Cambridge: Cambridge University Press.

GUMBEL, E. J. 1958: *The statistics of extremes*. New York: Columbia University Press.

Chapter 8

On integration in general

KNIGHT, B. and ADAMS, R. 1975 *Calculus I* and *Calculus II*. London: Unwin.

LANG, S. 1973: *A first course in calculus*. Reading, Massachusetts: Addison-Wesley.

For tables of standard integrals.

DWIGHT, H. B. 1957: *Tables of integrals and other mathematical data*. London: Macmillan.

Chapter 9

On general differential equations

MANOUIGAN, M. N. and NORTHCUTT, R. A. 1973: *Ordinary differential equations*. London: Charles E. Merrill.

STEPHENSON, G. 1973: *Mathematical methods for science students*. London: Longmans.

On complex numbers

FLANDERS, H. and PRICE, J. J. 1973: *Introductory college mathematics with linear algebra and finite mathematics*. New York: Academic Press.

References

DAVIDSON, D. A. 1973: Particle size and phospate analysis—evidence for the evolution of a tell. *Archaeometry* 15(1), 143–52.

DAVIDSON, D. A. 1977: The subdivision of a slope profile on the basis of soil properties: a case study from mid-Wales. *Earth Surface Processes* 2. 55–61.

DAVIDSON, D. A. 1978: *Science for physical geographers*. London: Edward Arnold.

DOBBIE, C. H. and WOLF, P. O. 1953: The Lynmouth flood of August 1952. *Proceedings of the Institute of Civil Engineers*.

DURY, G. H. 1959: Analysis of regional flood frequency in the Nene and Great Ouse. *Geographic Journal,* 125, 223–9.

EAST AFRICAN METEOROLOGICAL DEPARTMENT, 1964: *Climatological statistics for East Africa and the Seychelles, part III—Tanganyika and Zanzibar*.Nairobi, Kenya: EAMD.

FLANDERS, H. and PRICE, J. J. 1973: *Introductory college mathematics with linear algebra and finite mathematics* New York: Academic Press.

GABRIEL, K. R. and NEUMANN, J. 1962: A Markov Chain model for daily rainfall at Tel Aviv. *Quarterly Journal of the Royal Meteorological Society* 88, 90–95.

GARDINER, V. 1975: *Drainage basin morphometry*. British Geomorphological Research Group Technical Bulletin 14.

GRAY, J. M. 1974: The main rock platform of the Firth of Lom, western Scotland. *Institute of British Geographers, Transactions* 61, 81–100.

GREEN, J. F. N. 1935: The terraces of southernmost England. *Quarterly Journal of the Geological Society,* 92, 58–72.

GREGORY, K. and WALLING, D. 1973: *Drainage basin form and process*. London: Edward Arnold.

HACK, J. T. 1957: Studies of the longitudinal stream profiles in Virginia and Maryland. *US Geological Survey Professional Paper 294–B* 45–97.

HINDI, W. N. A. and KELWAY, P. S. 1977: Determination of storm velocities as an aid to the quality control of recording raingauge data. *Journal of Hydrology* 32, 115–37.

HORTON, R. E. 1945: Erosional development of streams and their drainage basins; hydrophysical approach to quantitative morphology. *Geological Society of America Bulletin* 56, 275–370.

HUDSON, N. W. 1971: *Soil Conservation*. London: Batsford.

HUFF, F. A. and NEILL, J. C. 1957: Areal representativeness of point rainfall. *Transactions of the American Geophysical Union* 38, 341–345.

KING, C. A. 1966: *Techniques in geomorphology*. London: Edward Arnold.

KIRKBY, M. J. 1977: Soil development models as a component of slope models. *Earth Surface Processes* **2**, 203–30.

KNIGHT, B. and ADAMS, R. 1975: *Calculus I* and *Calculus II*. London: Unwin.

LEOPOLD, L. B., WOLMAN, M. G. and MILLER, J. P. 1964: *Fluvial processes in geomorphology*. San Francisco: Freeman.

LINSLEY, R. K. and FRANZINI, J. B. 1972: *Water resources engineering*, second edition. New York: McGraw-Hill.

LINSLEY, R. K., KOHLER, M. A. and PAULUS, J. L. H. 1975: *Hydrology for engineers*. New York: McGraw-Hill.

MCCALLUM, D. 1959: The relationship between maximum rainfall intensity and time. *East African Meteorological Department*, Memoir III, No. 7, Nairobi.
MCINTOSH, D. H. and THOM, A. S. 1969: *Essentials of meteorology. Wykeham Science Series*. London: Wykeham.
MIAMI CONSERVATION DISTRICT, 1936: Storm rainfall of the eastern United States. *Technical Report part V*, 258–59.
MILLER, R. L. and KAHN, J. S. 1962: *Statistical analysis in the geological sciences*. New York: Wiley.
PANOFSKY, H. A. and BRIER G. W. 1958: Some applications of statistics and meteorology. Pennsylvania: Pennsylvania State University.
PEARSON, K. 1901: On the lines and planes of closest fit to systems of points in space. *Philosophical Magazine*, 6, 320–331.
PHILIPS, J. 1957–58: The theory of infiltration. *Soil Science*, **83–85**, various pages.
QUINN, F. H. 1977: Annual and seasonal flow variations through the Straits of Mackinac. *Water Resources Research*, 13, 137–44.
RAYNER, J. N. 1972: The application of harmonic and spectral analysis to the study of terrain. In *Spatial analysis in geomorphology*, 283–300, London: Macmillan.
RUHE, R. K. 1969: *Quaternary landscape in Iowa*. Ames, Iowa: Iowa State University Press.
SCHEIDEGGER, A. E. 1970: *Theoretical geomorphology*. Berlin: Springer-Verlag.
SHAW, J. and HEALY, T. R. 1977: Rectilinear slope formation in Antarctica. *Annals of the Association of American Geographers* 67, 46–54.
SMITH, K. 1972: *Water in Britain*. London: Macmillan.
STATHAM, I. 1976: A scree slope rockfall model. *Earth Surface Processes* **1**, 43–62.
STRAHLER, A. N. 1957: Quantitative analysis of watershed geomorphology. *Transactions of the American Geophysical Union* 38, 913–20.
SUMNER, G. N. 1977a: Sea breezes in hilly terrain. *Weather* **32**, 200–8.
SUMNER, G. N. 1977b: Sea breeze temperature and humidity contrasts at Lampeter, Dyfed. *Cambria* **4**, 187–98.
THORNES, J. E. 1971: state, environment and attribute in scree-slope studies. in Brunsden, D., editor, *Slopes, form and process*. Institute of British Geographers Special Publication 3.
WEYMAN, D. R. 1974: Runoff processes, contributing area and streamflow in a small upland catchment. *Institute of British Geographers Special Publication* **6**, 33–43.
YOUNG, A. 1972: *Slopes*. London: Longman.
YOUNGS, E. G. and AGGELIDES, S. 1976: Drainage to a water table analysed by the Green-Ampt approach. *Journal of Hydrology* **31**, 67–79.

Index